endorsed by edexcel

A Level Mathematics for Edexcel

Further Pure

FP1

Mark Rowland

OXFORD
UNIVERSITY PRESS

OXFORD
UNIVERSITY PRESS

Great Clarendon Street, Oxford OX2 6DP

Oxford University Press is a department of the University of Oxford.
It furthers the University's objective of excellence in research, scholarship, and education by publishing worldwide in

Oxford New York

Auckland Cape Town Dar es Salaam Hong Kong Karachi
Kuala Lumpur Madrid Melbourne Mexico City Nairobi
New Delhi Shanghai Taipei Toronto

With offices in

Argentina Austria Brazil Chile Czech Republic France Greece
Guatemala Hungary Italy Japan South Korea Poland Portugal
Singapore Switzerland Thailand Turkey Ukraine Vietnam

First published 2008

British Library Cataloguing in Publication Data

Data available

ISBN: 9780-19-911783 3

10 9 8 7 6 5 4

Printed in Great Britain by Ashford Colour Press Ltd, Gosport.

Paper used in the production of this book is a natural, recyclable product made from wood grown in sustainable forests. The manufacturing process conforms to the environmental regulations of the country of origin.

Acknowledgements

The photograph on the cover is reproduced courtesy of sharply-done/Fotolia

p58 /iStockphoto; **p78** /iStockphoto; **p104** /iStockphoto; **p146** /iStockphoto; **p154** /iStockphoto

The publishers would also like to thank John Rayneau and Ian Bettison for their expert help in compiling this book.

About this book

Endorsed by Edexcel, this book is designed to help you achieve your best possible grade in Edexcel GCE Further Mathematics Further Pure 1 unit.

Each chapter starts with a list of objectives and a 'Before you start' section to check that you are fully prepared. Chapters are structured into manageable sections, and there are certain features to look out for within each section:

Key points are highlighted in a blue panel.

Key words are highlighted in **bold blue** type.

Worked examples demonstrate the key skills and techniques you need to develop. These are shown in boxes and include prompts to guide you through the solutions.

EXAMPLE 2

Verify that $1 + 2i$ is a root of the equation $z^2 - 2z + 5 = 0$

Define $P(z) = z^2 - 2z + 5$

Substitute $z = 1 + 2i$:

$$P(1 + 2i) = (1 + 2i)^2 - 2(1 + 2i) + 5$$
$$= (-3 + 4i) - 2(1 + 2i) + 5$$
$$= 0 + 0i$$

Since $P(1 + 2i) = 0$, the complex number $1 + 2i$ is a root of the equation $z^2 - 2z + 5 = 0$

Derivations and additional information are shown in a panel.

Helpful hints are included as blue margin notes and sometimes as blue type within the main text.

Misconceptions are shown in the right margin to help you avoid making common mistakes.

Investigational hints prompt you to explore a concept further.

Each section includes an exercise with progressive questions, starting with basic practice and developing in difficulty. Some exercises also include 'stretch and challenge' questions marked with a stretch symbol .

At the end of each chapter there is 'Review' section which includes exam style questions as well as past exam paper questions. There are also two 'Revision' sections per unit which contain questions spanning a range of topics to give you plenty of realistic exam practice.

The final page of each chapter gives a summary of the key points, fully cross-referenced to aid revision. Also, a 'Links' feature provides an engaging insight into how the mathematics you are studying is relevant to real life.

At the end of the book you will find full solutions, a key word glossary, a list of formulae given in the Edexcel formulae booklet and an index.

Contents

FP1

1 Background knowledge

This chapter will remind you how to

- solve polynomial equations
- use trigonometric rules to solve problems
- sketch curves
- evaluate series
- differentiate a function to find an equation for the tangent and normal to its curve.

1.1 Solving polynomial equations

You can solve a quadratic equation by

1 factorising
2 completing the square
3 using the quadratic formula.

You will use the work in this section in Chapters 2 and 4.

If $ax^2 + bx + c = 0$, then $x = \dfrac{-b \pm \sqrt{b^2 - 4ac}}{2a}$

The discriminant, $b^2 - 4ac$, of a quadratic equation determines the nature of its roots.

$$\text{If } b^2 - 4ac \begin{cases} > 0 \text{ there are real, distinct roots} \\ = 0 \text{ there are real and equal roots} \\ < 0 \text{ there are no real roots.} \end{cases}$$

Refer to **C1** for revision of solving equations.

FP1

EXAMPLE 1

Solve the equation $3x^2 + 4x - 15 = 0$ by factorising.

$3x^2 + 4x - 15 = 0$

Factorise:

$(3x - 5)(x + 3) = 0$

Use intelligent guesswork: $-5 \times +3 = -15$

so $x = \frac{5}{3}$ or $x = -3$

EXAMPLE 2

Solve the equation $3x^2 + 4x - 15 = 0$ by completing the square.

$3x^2 + 4x - 15 = 0$

$3\left(x^2 + \frac{4}{3}x - 5\right) = 0$

Take out a factor of 3 to make the coefficient of x^2 equal to 1 inside the bracket.

Complete the square inside the bracket:

$$3\left(\left(x + \frac{2}{3}\right)^2 - \frac{49}{9}\right) = 0$$

$-\frac{4}{9} - 5 = -\frac{49}{9}$

Divide both sides by 3: $\left(x + \frac{2}{3}\right)^2 - \frac{49}{9} = 0$

Rearrange: $\left(x + \frac{2}{3}\right)^2 = \frac{49}{9}$

$$x = -\frac{2}{3} \pm \frac{7}{3}$$

$\sqrt{\frac{49}{9}} = \frac{\sqrt{49}}{\sqrt{9}} = \pm\frac{7}{3}$

$$x = \frac{5}{3} \quad \text{or} \quad x = -3$$

EXAMPLE 3

Use the quadratic formula to solve the equation
$3x^2 + 4x - 15 = 0$

$3x^2 + 4x - 15 = 0$ so $a = 3, b = 4, c = -15$

Write down the values of a, b and c.

Substitute into the formula:

$$x = \frac{-b \pm \sqrt{b^2 - 4ac}}{2a} = \frac{-4 \pm \sqrt{196}}{6}$$

$$= \frac{-4 \pm 14}{6}$$

$$x = \frac{5}{3} \quad \text{or} \quad -3$$

Discriminant
$b^2 - 4ac = 4^2 - 4 \times 3 \times (-15)$
$= 16 - (-180)$
$= 196 > 0$
Hence there are two distinct real roots.

The solutions of the equation $3x^2 + 4x - 15 = 0$ are $x = \frac{5}{3}$ or $x = -3$

You can solve a pair of simultaneous equations by eliminating one of the variables.

Refer to C1 for revision of simultaneous equations.

EXAMPLE 4

Solve the simultaneous equations

$2x - 3y = 6$ (1)
$xy = 12$ (2)

Make y the subject of equation (2):

$$xy = 12$$

$$y = \frac{12}{x}$$

Substitute for y in equation (1):

This eliminates y from equation (1).

$$2x - 3\left(\frac{12}{x}\right) = 6$$

Multiply all terms by x: $2x^2 - 36 = 6x$
Divide through by 2 to simplify: $x^2 - 18 = 3x$
Rearrange: $x^2 - 3x - 18 = 0$
Factorise: $(x - 6)(x + 3) = 0$

$$x = 6 \quad \text{or} \quad x = -3$$

Substitute each value of x into $y = \frac{12}{x}$ to find the corresponding value of y:

If $x = 6$, $y = \frac{12}{6} = 2$ and if $x = -3$, $y = \frac{12}{-3} = -4$

Hence the simultaneous equations have solutions
$x = 6, y = 2$ or $x = -3, y = -4$

FP1

You can solve a cubic equation $P(x) = 0$ by finding all the factors of the polynomial $P(x)$.

Refer to C2 for revision of cubic equations.

EXAMPLE 5

Solve $x^3 - 2x^2 - 3x + 6 = 0$
given that 2 is one root of the equation.

2 is a root of the equation $P(x) = 0$
where $P(x) = x^3 - 2x^2 - 3x + 6$
Hence $(x - 2)$ is a factor of $P(x)$.

Using the Factor Theorem – see C2.

Divide the factor $(x - 2)$ into $P(x)$ to find its other factors:

$$\begin{array}{r} x^2 + 0x - 3 \\ x - 2\,\overline{\big)\,x^3 - 2x^2 - 3x + 6} \\ \underline{x^3 - 2x^2} \\ -3x + 6 \\ \underline{-3x + 6} \\ 0 \end{array}$$

The quotient is $x^2 - 3$

Zero remainder confirms $(x - 2)$ is a factor of $P(x)$.

Hence $P(x) \equiv (x - 2)(x^2 - 3)$

Since $(x - 2)(x^2 - 3) = 0$
either $x - 2 = 0$ or $x^2 - 3 = 0$

Hence the equation $P(x) = 0$ has solutions
$x = 2$ or $x = \pm\sqrt{3}$.

FP1

Exercise 1.1

1 Solve these quadratic equations by factorising.

a $x^2 + 10x + 21 = 0$

b $x^2 - 14x + 13 = 0$

c $2x^2 + 7x - 15 = 0$

d $x^2 - 9x - 22 = 0$

2 Solve these quadratic equations by completing the square. Give your answers in surd form.

a $x^2 + 8x - 3 = 0$

b $x^2 - 5x + 2 = 0$

c $2x^2 + 5x - 1 = 0$

3 Solve these quadratic equations using the formula.

a $x^2 - 4x - 7 = 0$

b $2x^2 + 3x - 3 = 0$

c $3x^2 + 7x + 2 = 0$

4 Solve each of these quadratic equations using an appropriate method.
Give your answers in simplified surd form where necessary.

a $2x^2 - 11x + 12 = 0$

b $9x^2 + 12x - 8 = 0$

c $3x^2 - 4x = 2$

5 Use the method of elimination to solve each pair of simultaneous equations.

a $2x^2 + y^2 = 9$
$2x + y = 5$

b $x^2 - 3y^2 = 4$
$x - y = 8$

c $x^2 + y^2 = 6$
$x + y = 2$

d $y^2 = x$
$3x - y = 2$

6 Use the given root to find all the real roots of each cubic equation.

a $x^3 - 6x^2 + 5x + 12 = 0$ given that 3 is a root.

b $2x^3 + 5x^2 - x - 6 = 0$ given that −2 is a root.

c $x^3 - 12x + 16 = 0$ given that 2 is a root.

d $x^3 - 3x^2 + x + 5 = 0$ given that −1 is a root.

1.2 Trigonometric ratios and rules

You need to know the trigonometric ratios of these angles:

$\sin\frac{1}{6}\pi = \cos\frac{1}{3}\pi = \frac{1}{2}$ $\qquad \sin\frac{1}{3}\pi = \cos\frac{1}{6}\pi = \frac{\sqrt{3}}{2}$

$\sin\frac{1}{4}\pi = \cos\frac{1}{4}\pi = \frac{1}{\sqrt{2}}$ $\qquad \tan\frac{1}{6}\pi = \frac{1}{\sqrt{3}}, \tan\frac{1}{4}\pi = 1, \tan\frac{1}{3}\pi = \sqrt{3}$

Refer to C2 for revision on trigonometry.
This work will be continued in Chapters 2, 4 and 5.

You can solve trigonometric problems by using a combination of the sine rule and the cosine rule.

The cosine rule is given in the formula booklet (under C2).

For triangle ABC,

The sine rule gives $\frac{a}{\sin A} = \frac{b}{\sin B} = \frac{c}{\sin C}$

The cosine rule gives $c^2 = a^2 + b^2 - 2ab\cos C$

The area of a triangle is given by $A = \frac{1}{2}ab\sin C$

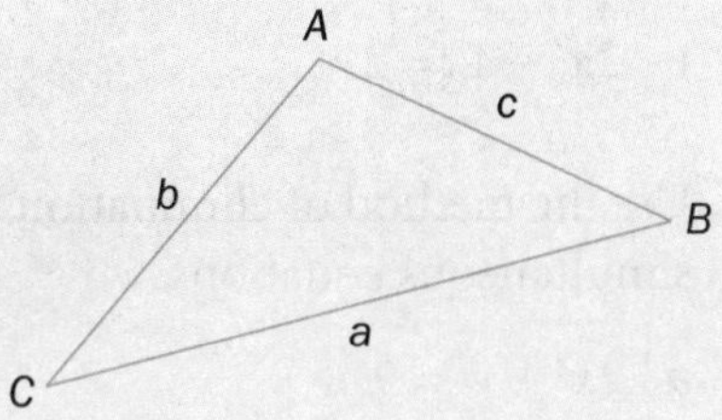

EXAMPLE 1

In the diagram, triangle ABC is such that $AC = 6$, $BC = 4\sqrt{3}$ and angle $ACB = \frac{1}{6}\pi$ radians.

a Find the exact length of AB.

b Find the exact area of triangle ABC.

c Determine whether or not triangle ABC is right-angled.

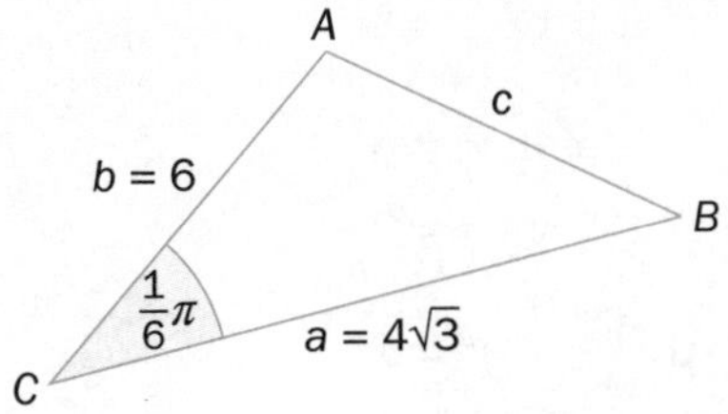

a Use the cosine rule to find the length of c:

$$c^2 = a^2 + b^2 - 2ab\cos C$$

$$= (4\sqrt{3})^2 + 6^2 - 2 \times 4\sqrt{3} \times 6 \times \frac{\sqrt{3}}{2}$$

$$= 48 + 36 - 72$$

$$= 12$$

$\cos C = \cos\frac{1}{6}\pi = \frac{\sqrt{3}}{2}$

Hence $c = \sqrt{12} = 2\sqrt{3}$

b Area $= \frac{1}{2}ab\sin C$

$$= \frac{1}{2} \times 4\sqrt{3} \times 6 \times \sin\frac{1}{6}\pi$$

$$= 6\sqrt{3}$$

$\sin C = \sin\frac{1}{6}\pi = \frac{1}{2}$

Hence the triangle has area $6\sqrt{3}$ square units.

Part **c** is answered on the next page.

c Check if the triangle obeys Pythagoras' Theorem:

Find the sum of the squares on the two shorter sides:

$b^2 + c^2 = 6^2 + (2\sqrt{3})^2 = 36 + 12 = 48$

$6 < 4\sqrt{3}$ and $2\sqrt{3} < 4\sqrt{3}$

and $a^2 = (4\sqrt{3})^2 = 48$

Sum of the squares on the two shorter sides equals the square on the hypotenuse.

Hence $a^2 = b^2 + c^2$ and so the triangle is right-angled. The right angle is at A.

You could calculate angle A directly using the sine or cosine rule.

Exercise 1.2

1 Find the exact value of these, giving answers in simplified surd form where appropriate.

a $\sin\frac{1}{6}\pi + \cos\frac{1}{6}\pi$

b $\sin^2\frac{1}{4}\pi + \cos\frac{1}{3}\pi$

c $\tan\frac{1}{3}\pi - \tan\frac{1}{6}\pi$

d $\sin\frac{1}{6}\pi\cos\frac{1}{6}\pi\tan\frac{1}{6}\pi$

2 If $x = \arcsin\left(\frac{\sqrt{3}}{2}\right)$ find the value of $\tan^2 x$.

Draw a diagram.

$\arcsin\left(\frac{\sqrt{3}}{2}\right) = \sin^{-1}\left(\frac{\sqrt{3}}{2}\right)$

3 The diagram shows triangle ABC, where $AB = 2$, $BC = \sqrt{6}$ and angle $CAB = \frac{2}{3}\pi$ radians.

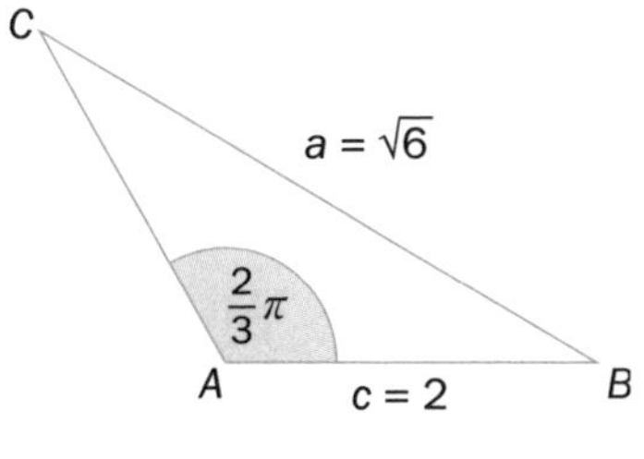

a Find, in exact form, angles C and B.

b Show that $AC = \sqrt{3} - 1$ and hence find the exact area of triangle ABC.

1.3 Sketching curves

Polynomial curves

You need to be able to sketch the general shape of a curve by inspecting its equation.

You will need to be able to sketch curves for the work covered in Chapter 3.

EXAMPLE 1

Sketch the graph of the curve with equation $y = 2x^2 - x - 3$

Find the roots of the equation:

$$2x^2 - x - 3 = 0$$

$$(2x - 3)(x + 1) = 0$$

$$x = \frac{3}{2} \quad \text{or} \quad -1$$

The y-intercept is –3.

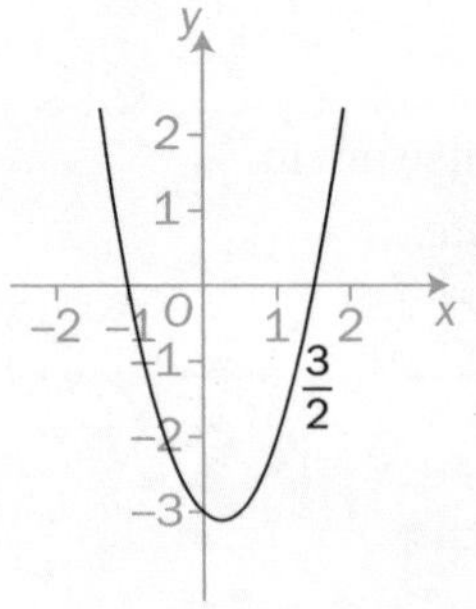

See C1 for revision on sketching quadratics.

Label the roots and y-intercept on your sketch.

Quadratic curves with positive x^2 coefficient are ∪–shaped.

Exponential curves

You should be familiar with the shape of the graph with equation $y = a^x$ for $a > 0$.

See C2 for revision of the function a^x.

You will use exponential graphs in Chapter 3.

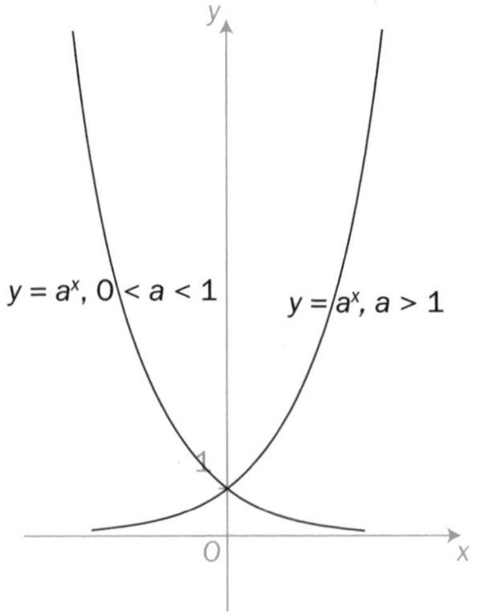

EXAMPLE 2

a Sketch, on the same diagram, the curves with equations $y = x - 1$ and $y = 2^{-x}$

b Hence show that the equation $2^{-x} + 1 - x = 0$ has exactly one real root.

---

a

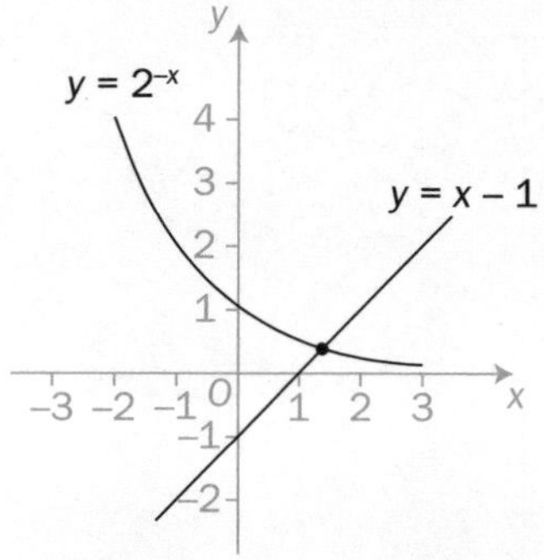

$y = 2^{-x} = \frac{1}{2^x} = \left(\frac{1}{2}\right)^x$

b The roots of the equation $2^{-x} + 1 - x = 0$ correspond to the points of intersection of the graphs with equations $y = x - 1$ and $y = 2^{-x}$.

Since the sketch in part **a** shows only one point of intersection, the equation $2^{-x} + 1 - x = 0$ has exactly one real root.

Intersection points of graphs $y = f(x)$ and $y = g(x)$ correspond to the roots of the equation $f(x) - g(x) = 0$ – refer to C1

FP1

Exercise 1.3

1 Sketch, on separate diagrams, the curves with these equations.

a $y = 3x^2 - 10x - 8$

b $y = 2x^2 - 2x + 1$

c $y = (x + 2)(x - 3)(1 - x)$

2 **a** On the same diagram sketch the graphs of these equations.

i $y = 2^x$
ii $y = 9 - x^2$

b Hence state the number of roots of the equation $2^x + x^2 - 9 = 0$

Sequences and series

You can work out terms of a sequence defined by an iterative formula.

An iterative formula is also called a recurrence relation – see C1

You will use this knowledge in Chapters 3 and 7.

EXAMPLE 1

Find the second and third terms of the sequence defined by the iterative formula

$u_{n+1} = 2 - 3u_n,\ u_1 = 2$

The first term $u_1 = 2$ is given.

$u_2 = 2 - 3u_1$
$= 2 - 3 \times 2$
$= -4$

$u_3 = 2 - 3u_2$
$= 2 - 3 \times (-4)$
$= 14$

You can use the ANS key on your calculator to generate these values.

Hence the second and third terms of this sequence are −4 and 14.

You can express a series using Σ notation.

A series is the sum of the terms of a sequence.

A series is arithmetic if the terms in its sum form an arithmetic sequence.

If the series is arithmetic then you can use the formula

$$S_n = \frac{1}{2}n[2a + (n-1)d]$$

to calculate the sum S_n of the first n terms of the series.

a is the first term,
d is the common difference.

EXAMPLE 2

Find the sum of the series $\sum_{r=1}^{20}(1.5r + 4)$

You will continue working with series in Chapters 6 and 7.

$$\sum_{r=1}^{20}(1.5r + 4) = (1.5 \times 1 + 4) + (1.5 \times 2 + 4) + \ldots + (1.5 \times 20 + 4)$$
$$= 5.5 + 7 + \ldots + 34$$

The series is arithmetic with $n = 20$ terms.
The first term, a, is 5.5 and the common difference, d, is 1.5.

$d = 7 - 5.5 = 1.5$

Use the formula $S_n = \frac{1}{2}n[2a + (n-1)d]$ where $n = 20$, $a = 5.5$ and $d = 1.5$:

$$\sum_{r=1}^{20}(1.5r + 4) = S_{20} = \frac{1}{2} \times 20\,(2 \times 5.5 + 19 \times 1.5) = 395$$

Exercise 1.4

1 Use the iterative formula to calculate the terms u_2, u_3 and u_4 in each case.

a $u_{n+1} = 3 + 2u_n,\ u_1 = -1$

b $u_{n+1} = u_n - \dfrac{1}{u_n},\ u_1 = 2$

c $u_{n+1} = 1 + \dfrac{2u_n}{(u_n + 1)},\ u_1 = -3$

d $u_{n+1} = 2\left(u_n + \dfrac{1}{\sqrt{u_n}}\right),\ u_1 = 1$

2 Evaluate each of these arithmetic series.

a $\sum_{r=1}^{15}(2r - 1)$

b $\sum_{r=1}^{100}(4 - 0.1r)$

c $\sum_{r=6}^{30}\left(\frac{1}{2}r + 1\right)$

d $\sum_{r=1}^{25}\dfrac{r^2 - 1}{r + 1}$

3 A sequence is defined by the iterative formula

$$u_{n+1} = u_n - \frac{1}{(2u_n + 1)}, \quad u_1 = -1$$

a Calculate the values of the iterates u_2, u_3 and u_4.

b Hence state the value of

i u_{32}

ii u_{1001}

iii $u_{n(n+1)}$ for $n \geqslant 1$

FP1

1.5 Differentiation, tangents and normals

You can **differentiate** a curve to find the gradient of that curve at any point.

You will use the work in this section in Chapters 3 and 4.

EXAMPLE 1

Find the gradient of the curve with equation $y = x^3 - 4x + 1$ at the point $P(2, 1)$.

Differentiate the equation of the curve:

Refer to C1 for differentiating powers of x.

$$y = x^3 - 4x + 1 \quad \text{so} \quad \frac{dy}{dx} = 3x^2 - 4$$

Substitute the x-coordinate of P into $\frac{dy}{dx}$:

At P, $x = 2$

$$\frac{dy}{dx} = 3 \times 2^2 - 4 = 8$$

The gradient of the curve at P is 8.

You can use the notation $f'(x)$ to represent the gradient of the curve with equation $y = f(x)$ at the point $P(x, y)$.

Refer to C1 for $f'(x)$ notation.

Once you know the gradient of a curve at a given point, you can find the equation of the tangent and the normal to the curve at that point.

Refer to C1 for the tangent and normal to a curve.

EXAMPLE 2

A curve has equation $y = f(x)$, where $f(x) = 4x + \frac{1}{x^2}$, $x > 0$

Find the equations of the tangent and the normal to this curve at the point $P(1, 5)$.

Express $f(x)$ in a form appropriate for differentiation:

$$f(x) = 4x + x^{-2}$$

Differentiate:

$$f'(x) = 4 - 2x^{-3}$$

The derivative of x^n is nx^{n-1}.

Substitute $x = 1$ into $f'(x)$ to find $f'(1)$:

At $P(1, 5)$, $x = 1$

$$f'(1) = 4 - 2 \times 1^{-3} = 2$$

The tangent, T, to this curve at point $P(1, 5)$ has gradient 2

Hence an equation for T is $y - 5 = 2(x - 1)$

that is $y = 2x + 3$

The gradient of the normal, N, to this curve at $P(1, 5)$ is $-\frac{1}{2}$

Refer to C1 – for perpendicular lines, product of gradients $= -1$.

Hence an equation for N is $y - 5 = -\frac{1}{2}(x - 1)$

that is $2y + x = 11$

Exercise 1.5

For part **f**, expand and simplify first.

1 Find an equation for the tangent and the normal to each curve at the given point.

a $y = x^2 - x + 2 \quad P(2, 4)$

b $y = 2x^3 - 4x - 1 \quad P(-1, 1)$

c $y = x^2 - \frac{1}{x} \quad P(1, 0)$

d $y = x^2 - 6\sqrt{x} - 1$ at the point where $x = 4$

e $y = \sqrt[3]{x} - \frac{16}{x}$ at the point where $x = -8$

f $y = (2x - 1)(x + 2) \quad P(1, 3)$

2 **a** Show that $\frac{2x^2 + 1}{x^3} \equiv 2x^{-1} + x^{-3}$

b Hence find the gradient of the curve with equation

$$y = \frac{2x^2 + 1}{x^3}$$ at the point where $x = 1$

3 Find $f'(x)$ for each function.

a $f(x) = 2\sqrt{3x}$

b $f(x) = \frac{1 - x^2}{x^4}$

c $f(x) = \frac{x^3 + x^2 + x + 1}{x}$

4 A curve has equation $y = f(x)$ where $f(x) = (x^2 - 8)(x + 1)$

a Show that $f'(x) = 3x^2 + 2x - 8$

b Find an equation for the normal to this curve at the point where $x = -3$. Give your answer in the form $ay + bx + c = 0$, for integers a, b and c to be stated.

c Find the x-coordinates of the stationary points on this curve.

5 $f(x) = 2x^3 + x - 1$

a Find an equation for the tangent to the curve $y = f(x)$ at the point where $x = 1$.

b Find the coordinates of the points on this curve at which the gradient is 2.5.

c Show that $f'(x) > 0$ for all real values of x and state the value of x at which the gradient of the curve $y = f(x)$ is at its minimum.

d Sketch the curve with equation $y = f(x)$

You do not need to find the roots of this graph.

1

Exit

Summary

Refer to

- The quadratic formula states that

  if $ax^2 + bx + c = 0$ then $x = \dfrac{-b \pm \sqrt{b^2 - 4ac}}{2a}$ 1.1

- $b^2 - 4ac$ is the discriminant of the equation $ax^2 + bx + c = 0$

  If $b^2 - 4ac \begin{cases} > 0 & \text{there are real distinct roots} \\ = 0 & \text{there are real and equal roots} \\ < 0 & \text{there are no real roots.} \end{cases}$

- For triangle ABC

  The sine rule states that $\dfrac{a}{\sin A} = \dfrac{b}{\sin B} = \dfrac{c}{\sin C}$

  The cosine rule states that $c^2 = a^2 + b^2 - 2ab \cos C$

  The area, A, of the triangle is given by

  $A = \frac{1}{2} ab \sin C$ 1.2

- For an arithmetic series with first term a and

  common difference d

  $S_n = \frac{1}{2} n[2a + (n-1)d]$

  where S_n is the sum of the first n terms of the series. 1.4

- For a function $f(x) = ax^n$,

  $f'(x) = an\,x^{n-1}$ 1.5

FP1

2 Complex numbers

This chapter will show you how to:

- use complex numbers to solve equations such as $3x^2 - 2x + 1 = 0$
- add, subtract, multiply and divide complex numbers
- represent complex numbers geometrically
- solve polynomial equations of up to degree 4, such as $x^4 + 3x^3 - 12x - 16 = 0$

Refer to **1.1, 1.2**

Before you start

You should know how to:

1 Solve a quadratic equation.

e.g. Solve $2x^2 + x - 10 = 0$

$2x^2 + x - 10 = (2x + 5)(x - 2)$

Since $2x^2 + x - 10 = 0$, $(2x + 5)(x - 2) = 0$

hence $x = -\frac{5}{2}, 2$

2 Solve a pair of simultaneous equations.

e.g. Solve the simultaneous equations

$x^2 + 2y^2 = 6, x - y = 3$

$x^2 + 2y^2 = 6$ and $x = y + 3$

so $(3 + y)^2 + 2y^2 - 6 = 0$

$3y^2 + 6y + 3 = 0$

$3(y + 1)^2 = 0$

Hence $y = -1, x = 2$

3 Solve a polynomial equation.

e.g. Solve the equation $x^3 - 2x^2 - x + 2 = 0$ given that $x = 2$ is one of its roots.

$$\begin{array}{r|l} & x^2 \quad\quad - 1 \\ x - 2 & x^3 - 2x^2 - x + 2 \\ & x^3 - 2x^2 \\ & \quad\quad\quad - x + 2 \\ & \quad\quad\quad - x + 2 \\ & \quad\quad\quad\quad\quad 0 \end{array}$$

$x^3 - 2x^2 - x + 2 = 0$ so $(x - 2)(x + 1)(x - 1) = 0$

Hence $x = 2$ or $x = \pm 1$

Check in:

1 Solve

a $x^2 + x - 12 = 0$

b $9x^2 - 6x = 0$

c $4x^2 - 4x + 1 = 0$

See C1 for revision.

2 Solve each pair of simultaneous equations.

See C1 for revision.

a $2x^2 + y^2 = 9$
$x + y = 3$

b $2x^2 + 2y^2 = 1$
$4xy = 1$

c $x^4 - y^4 = 15$
$x^2 - y^2 = 3$

3 Solve the equation

$3x^3 - 5x^2 - 4x + 4 = 0$

given that $x = -1$ is one of its roots.

2.1 The imaginary number i

The square root of any number, $x \geqslant 0$, is a real number,

but what is the square root of a negative number such as -9?

Real numbers lie on the number line.

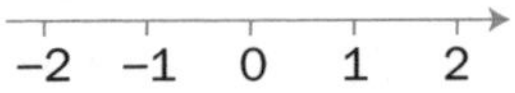

Consider the equation

$$x^2 + 9 = 0$$

then

$$x^2 = -9$$

$$x = \sqrt{-9}$$

$$= \sqrt{9 \times -1}$$

$$= \sqrt{9} \times \sqrt{-1}$$

since $\sqrt{ab} = \sqrt{a} \times \sqrt{b}$

$$= \pm 3 \times \sqrt{-1}$$

You can write the square root of any negative number as the product of a real number and $\sqrt{-1}$.

■ $\sqrt{-1}$ is an **imaginary number** and is denoted by the letter i.

So $\sqrt{-9} = \pm 3i$

and the solution to the equation $x^2 + 9 = 0$ is $x = \pm 3i$.

$x = \pm 3i$ are the roots of the equation $x^2 + 9 = 0$

You can check each answer is correct by substituting it into the given equation.

When $x = 3i$, $x^2 + 9 = (3i)^2 + 9$

$$= 9i^2 + 9$$

$$= -9 + 9$$

$$= 0 \quad \text{as required}$$

$(3i)^2 = 3i \times 3i = 9i^2$

$9i^2 = 9 \times -1 = -9$

Use this method to check that $x = -3i$ is also a root of this equation.

EXAMPLE 1

Find $\sqrt{-25}$, $\sqrt{-12}$ and $\sqrt{-7}$.

$\sqrt{-25} = \pm\sqrt{25}i = \pm 5i$

$\sqrt{-12} = \pm\sqrt{12}i = \pm\sqrt{4} \times \sqrt{3}i = \pm 2\sqrt{3}i$

$\sqrt{-7} = \pm\sqrt{7}i$

Exercise 2.1

1 Express each square root in terms of the imaginary number i.

a $\sqrt{-16}$ b $\sqrt{-36}$

c $\sqrt{-100}$ d $\sqrt{-20}$

2 Solve these equations, giving your answers in terms of the imaginary number i.
Check each answer by substitution.

a $x^2+4=0$ b $x^2+49=0$

c $x^2+225=0$ d $2x^2+50=0$

e $3x^2+27=0$ f $4x^2+1=0$

g $9x^2+4=0$ h $27x^2+6=0$

3 Solve these equations, giving your answers in simplified surd form.

a $x^2+8=0$ b $x^2+18=0$

c $3x^2+36=0$ d $4x^2+180=0$

e $5x^2+120=0$ f $\frac{1}{2}x^2+14=0$

g $12x^2+1=0$ h $\frac{1}{3}x^2+\frac{1}{4}=0$

4 Solve each equation, giving answers in terms of the positive number a.

a $x^2+a^2=0$ b $x^2+9a^2=0$

c $x^2+a^4=0$ d $(ax)^2+1=0$

5 The equation $ax^2+b=0$, where a and b are positive integers, has a root $\frac{5\sqrt{2}}{2}i$.

a Show that $b=\frac{25}{2}a$

b Given that $b<50$, find the value of a and the value of b.

6 Solve these equations.

a $x^2+\sqrt{2}=0$ b $x^4=16$

c $\frac{1}{x^2}+9=0$ d $\frac{2x^2+25}{x^2}=1$

FP1

2.2 Complex numbers

You can use the imaginary number i to solve a quadratic equation.

EXAMPLE 1

Solve the equation $x^2 - 6x + 13 = 0$.

If $x^2 - 6x + 13 = 0$, $\quad x = \dfrac{-(-6) \pm \sqrt{(-6)^2 - 4 \times 1 \times 13}}{2 \times 1}$

Using the quadratic formula with $a = 1$, $b = -6$, $c = 13$. See C1 for revision.

$= \dfrac{6 \pm \sqrt{-16}}{2}$

Discriminant $= 36 - 52 = -16$

$= \dfrac{6 \pm 4i}{2}$

$\sqrt{-16} = \sqrt{16} \times \sqrt{-1} = \pm 4i$

Divide by 2: $= 3 \pm 2i$

The solutions to the equation $x^2 - 6x + 13 = 0$ are $x = 3 \pm 2i$

$3 + 2i$ is an example of a complex number.

An expression of the form $a + ib$ where a and b are *real* numbers, is called a complex number.

You write $a, b \in \mathbb{R}$ for 'a and b are real numbers'.

You can use the letter z to denote a complex number.

$z = a + ib$

z is a complex number

b is the imaginary part of z
$b = \text{Im}(z)$

a is the real part of z
$a = \text{Re}(z)$

You can write the equation in Example 1 using the letter z instead of x.
The equation $z^2 - 6z + 13 = 0$ has solutions $z = 3 \pm 2i$

You need to be able to identify the real and imaginary parts of a complex number.

EXAMPLE 2

Find the real and imaginary parts of

a $z = 5 - 2i$ **b** $w = \dfrac{-1 + 3i}{2}$

You can use other letters, such as w, to denote a complex number.

a Compare $5 - 2i$ with $a + ib$:

$a = 5, b = -2$

so $\text{Re}(z) = 5, \text{Im}(z) = -2$

b Express $\dfrac{-1 + 3i}{2}$ as $-\dfrac{1}{2} + \dfrac{3}{2}i$:

so $\text{Re}(w) = -\dfrac{1}{2}$ and $\text{Im}(w) = \dfrac{3}{2}$

You can divide the real and imaginary parts by the denominator.

A complex number with real part zero is a **purely imaginary number.**

e.g. The number $z = 0 + 3i$ is purely imaginary
and is abbreviated to $z = 3i$.

A complex number with zero imaginary part is a **real number.**

e.g. $z = 3 + 0i$ is treated as the real number 3
and is abbreviated to $z = 3$.

The zero complex number $z = 0 + 0i$, has real and imaginary parts both zero.
This number is abbreviated to $z = 0$.

The quadratic equation $ax^2 + bx + c = 0$ has complex roots if the discriminant $b^2 - 4ac < 0$.

See C1 for revision of the discriminant.

EXAMPLE 3

By finding the value of the discriminant state the nature of the roots of these equations.

a $x^2 + 4x - 5 = 0$

b $x^2 - 4x + 5 = 0$

c $x^2 + 2x + 1 = 0$

---

a $x^2 + 4x - 5 = 0$

$a = 1, b = 4, c = -5$

$b^2 - 4ac = 16 + 20$

$= 36$

$\therefore x^2 + 4x - 5 = 0$ has two real roots.

b $x^2 - 4x + 5 = 0$

$a = 1, b = -4, c = 5$

$b^2 - 4ac = 16 - 20$

$= -4$

$\therefore x^2 - 4x + 5 = 0$ has two complex roots.

c $x^2 + 2x + 1 = 0$

$a = 1, b = 2, c = 1$

$b^2 - 4ac = 4 - 4$

$= 0$

$\therefore x^2 + 2x + 1 = 0$ has one real root.

Exercise 2.2

1 Write down the real and imaginary parts of each complex number.

a $z = 3 + 2i$ b $z = 4 - 5i$

c $z = -1 + 4i$ d $z = 2 - 6i$

e $z = \frac{1}{2} + \frac{i}{3}$ f $z = -5 - 3i$

g $z = -7$ h $z = \sqrt{2}$

i $z = \sqrt{-4}$

2 Find the real and imaginary parts of each complex number.
Give your answers in simplified surd form where appropriate.

a $z = \frac{4 - 6i}{2}$ b $z = \frac{-8 - 4i}{2}$

c $z = \frac{i}{3}$ d $z = \frac{2 + 4i}{6}$

e $z = \frac{\sqrt{8} - 4i}{2}$ f $z = \frac{3 + \sqrt{6}i}{\sqrt{3}}$

FP1

3 a Find all complex numbers of the form $z = a + a^2 i$ for $a \in \mathbb{R}$, with imaginary part 9.

b Find all complex numbers of the form $z = 2k + k^2 i$ for $k \in \mathbb{R}$ such that $\text{Re}(z) = \text{Im}(z)$.

4 By finding the discriminant state the nature of the roots of these equations.

a $x^2 - 3x + 5 = 0$ b $x^2 + 3x + 1 = 0$

c $2x^2 + x - 1 = 0$ d $4x^2 - 4x + 1 = 0$

e $4x^2 + 7x + 5 = 0$

5 Solve these equations.
Give your answers in the form $a + ib$ where $a, b \in \mathbb{R}$.

a $z^2 + 4z + 5 = 0$ b $z^2 - 2z + 5 = 0$

c $z^2 - 4z + 13 = 0$ d $2z^2 + 2z + 5 = 0$

e $5z^2 + 2 = 2z$ f $8z^2 + 17 = 12z$

For exact answers, leave in surd form.

6 Solve these equations.
Give each answer in the form $a + ib$, where $a, b \in \mathbb{R}$ are exact.

a $z^2 + 3z + 3 = 0$ **b** $2z^2 - z + 1 = 0$

c $\frac{1}{2}z^2 + 3 + 2z = 0$ **d** $4z^2 + 3 = 5z$

e $3(z^2 + 1) = 5z$ **f** $z(11 - 5z) = 7$

7 Find, in simplified surd form, the solutions to each equation.

a $z^2 - z + 1 = 0$ **b** $3z^2 - 2z + 3 = 0$

c $2z^2 + 4z + 3 = 0$ **d** $z^2 + 3z + 9 = 0$

e $2z^2 + 4z + 5 = 0$ **f** $9z^2 - 6z + 4 = 0$

8 Solve each equation. Express the imaginary part of each solution as an integer multiple of $\sqrt{3}$.

a $z^2 - 2z + 4 = 0$ **b** $z^2 + 6z + 12 = 0$

c $z^2 - 6z + 21 = 0$ **d** $z^2 + 2z + 49 = 0$

9 Solve these equations. Give answers in simplified surd form where appropriate.

a $z(z - 3) = -5$ **b** $2z(1 - z) = 7$

c $z(3z - 1) = z - 5$ **d** $\frac{4}{z} = 3 - z$

e $\frac{3z}{z^2 + 3} = 1$ **f** $2z + \frac{7}{z} = 4$

10 **a** Show that $(z - 1)(z^2 + z + 1) \equiv z^3 - 1$

b Hence solve the equation $z^3 = 1$

FP1

2.3 The arithmetic of complex numbers

You can add, subtract, multiply and divide complex numbers.

You can apply many of the rules for ordinary arithmetic to complex numbers.

When calculating with complex numbers always replace i^2 with -1.

EXAMPLE 1

If $z = 2 + 3i$ and $w = 5 - 2i$ find

a $z + w$ **b** $z - w$ **c** zw

This example shows you how to add, subtract and multiply two complex numbers, z and w.

a $z + w = (2 + 3i) + (5 - 2i)$
$= (2 + 5) + i(3 - 2)$
$= 7 + i$

Collect the real and imaginary parts separately.

b $z - w = (2 + 3i) - (5 - 2i)$
$= 2 + 3i - 5 + 2i$
$= (2 - 5) + i(3 + 2)$
$= -3 + 5i$

$-(5 - 2i) = -5 - (-2)i$

You may find it helpful to use brackets when subtracting.

c $zw = (2 + 3i)(5 - 2i)$
$= 10 - 4i + 15i - 6i^2$
$= 10 + 11i + 6$
$= 16 + 11i$

Expand the brackets in the usual way.

$-6i^2 = -6 \times -1 = 6$

When multiplying a complex number by a number k, multiply its real part by k and its imaginary part by k.

e.g. $3(4 - i) = (3 \times 4) + (3 \times -i)$
$= 12 - 3i$

and $2i(3 + 5i) = 6i + 10i^2$
$= -10 + 6i$

$2i \times 5i = 10i^2 = -10$

The product of z with itself is written as z^2. Similarly, z^3 means $z \times z \times z$.

$z^2 = z \times z$, as in normal algebra.

e.g. If $z = 3 + i$ then $2z^2 = 2(3 + i)^2$
$= 2(9 + 6i - 1)$
$= 16 + 12i$

$i^2 = -1$

You can solve simultaneous equations involving complex numbers.

EXAMPLE 2

The complex numbers z and w satisfy the simultaneous equations

$$3z + w = 11 - 10i \quad (1)$$
$$z + w = 5 - 2i \quad (2)$$

Solve the equations to find z and w, giving each answer in the form $a + ib$ for $a, b \in \mathbb{R}$.

Subtract (2) from (1) to eliminate w:

$$3z + w - z - w = 11 - 10i - (5 - 2i)$$
$$2z = 11 - 10i - 5 + 2i$$

Collect the real parts and imaginary parts separately.

$$2z = 6 - 8i$$
$$z = \frac{6 - 8i}{2}$$
$$z = 3 - 4i$$

Simplify the real and imaginary parts as much as possible.

Substitute $z = 3 - 4i$ into (2) to find w:

(2) is easier to work with than (1).

$$z + w = 5 - 2i$$
$$(3 - 4i) + w = 5 - 2i$$
$$w = (5 - 2i) - (3 - 4i)$$
$$= 2 + 2i$$

$(5 - 2i) - (3 - 4i)$
$= 5 - 2i - 3 + 4i$
$= 2 + 2i$

Hence $z = 3 - 4i$ and $w = 2 + 2i$

FP1

You can divide a complex number, z, by another complex number, w, provided $w \neq 0 + 0i$.

An expression such as $\frac{z}{w}$ is called a **quotient**.

EXAMPLE 3

Find $\frac{5}{2 + 3i}$ in the form $a + bi$, where a and b are real numbers.

You need to transform the divisor into a real number.

The divisor is $2 + 3i$

Make the denominator real by multiplying both the top and the bottom of the fraction by $2 - 3i$:

$2 - 3i$ is the **complex conjugate** of $2 + 3i$. See Section 2.4.

Compare with rationalising $\frac{5}{2 + \sqrt{3}}$

See C1 for revision.

$$\frac{5}{2 + 3i} = \frac{5}{2 + 3i} \times \left(\frac{2 - 3i}{2 - 3i}\right)$$

$$= \frac{5(2 - 3i)}{(2 + 3i)(2 - 3i)}$$

$$= \frac{10 - 15i}{13}$$

$(2 + 3i)(2 - 3i) = 4 - 6i + 6i - 9i^2$
$= 4 + 9$
$= 13$, a real number

Hence $\frac{5}{2 + 3i} = \frac{10}{13} - \frac{15}{13}i$

EXAMPLE 4

Find the quotient $\frac{1+2i}{3-4i}$ in the form $a+bi$, where $a, b \in \mathbb{R}$.

Multiply both the top and the bottom of the fraction by $3 + 4i$ to eliminate the complex divisor:

$$\frac{1+2i}{3-4i} = \frac{(1+2i)(3+4i)}{(3-4i)(3+4i)}$$

$$= \frac{3+4i+6i+8i^2}{9+16}$$

$$= \frac{3+10i-8}{25}$$

$$= \frac{-5+10i}{25}$$

Hence $\frac{1+2i}{3-4i} = -\frac{1}{5} + \frac{2}{5}i$

$(3-4i)(3+4i)$
$= 9 + 12i - 12i - 16i^2$
$= 9 + 16$

$8i^2 = 8 \times (-1) = -8$

Divide the real and imaginary parts by 25 and separate them.

You can use division to solve simultaneous equations involving complex numbers.

EXAMPLE 5

The complex numbers z and w satisfy the simultaneous equations

$$3z + w = 6 + i \quad (1)$$

$$z + iw = 3 + 4i \quad (2)$$

Solve the equations to find z and w, giving each answer in the form $a + ib$ for $a, b \in \mathbb{R}$.

Multiply (2) by 3: $3z + 3iw = 9 + 12i$ (3)

Subtract (1) from (3): $3z + 3iw - 3z - w = 9 + 12i - 6 - i$

$$(-1+3i)w = 3 + 11i$$

Hence $w = \frac{3+11i}{-1+3i}$

To express w in the required form, multiply the top and bottom of $\frac{3+11i}{-1+3i}$ by $(-1-3i)$:

$$w = \frac{3+11i}{-1+3i} = \frac{(3+11i)(-1-3i)}{(-1+3i)(-1-3i)}$$

$$= \frac{30-20i}{10}$$

$$= 3 - 2i$$

$(3+11i)(-1-3i)$
$= -3 - 9i - 11i - 33i^2$
$= 30 - 20i$
$(-1+3i)(-1-3i)$
$= 1 + 3i - 3i - 9i^2$
$= 1 + 9 = 10$

Substitute $w = 3 - 2i$ into equation (1) to find z:

$$3z + w = 6 + i \quad (1)$$

$$3z + (3-2i) = 6 + i$$

$$3z = 3 + 3i$$

$$z = 1 + i$$

Hence $z = 1 + i$ and $w = 3 - 2i$

(1) is easier to use to find z than (2).

$(6+i) - (3-2i) = 6 + i - 3 + 2i$
$= 3 + 3i$

Divide the real and imaginary parts of $(3+3i)$ by 3.

FP1

Exercise 2.3

Unless indicated otherwise, give answers in the form $a + ib$ where $a, b \in \mathbb{R}$.

1 Simplify

a $(3 + 2i) + (4 + 3i)$ **b** $(8 - i) - (4 + 2i)$ **c** $(-4 - 2i) + (-2 - 3i)$

d $(3 - 6i) - (2 - 6i)$ **e** $3(2 + 3i)$ **f** $3i(1 + 4i)$

g $4 + (5 + 3i)i$ **h** $3(1 - 2i) + i(2 + i)$ **i** $i(2 + 3(2i + 1))$

2 By making z the subject, or otherwise, solve these equations.

a $4z - 3 = 5 + 4i$ **b** $4z - 3i = 4 + 3i$ **c** $6z + 5 = i(3 - 5i)$

3 Solve each pair of simultaneous equations.

a $3z + 2w = -3 + 11i$, $z + w = 1 + 4i$ **b** $2z + 3w = 2 + 11i$, $3z - w = 3$ **c** $4z + 3w = 5 + 2i$, $3z - 2w = 8 - 7i$

4 Simplify

a $(-2 + 3i)(4 + 2i)$ **b** $(-5 + 3i)(2 - 4i)$ **c** $(-2 + i)^2$

FP1

5 Find the following quotients.
Give each answer in the form $a + bi$ where $a, b \in \mathbb{R}$ are exact.

a $\dfrac{5 + 3i}{2 + i}$ **b** $\dfrac{2 - i}{1 - 3i}$ **c** $\dfrac{3 - 4i}{2i}$

6 By making z the subject, or otherwise, solve these equations.

a $(3 + i)\, z = 3 + 11i$ **b** $\dfrac{7 - 17i}{z + 1} = 2 - 3i$ **c** $\dfrac{i}{z - i} = 1 + i$

d $\dfrac{2 + 3i}{i(z + i)} = 3 - 2i$ **e** $\dfrac{z}{z + i} = 3i$ **f** $\dfrac{2z - i}{iz + 2} = 1$

7 Solve these pairs of simultaneous equations.

a $z + w = 2 + 4i$, $z + iw = 1 - i$ **b** $3z + w = -9 + 8i$, $z + iw = 7i$ **c** $2z + 3w = 20 - 7i$, $iz - 2w = -6 + 6i$

8 Solve the equation $\dfrac{1}{z - 1} + \dfrac{1}{z + 1} = \dfrac{1}{z}$

9 Solve each pair of simultaneous equations.
For part **c** consider the expansion $(z + iw)^2$

a $z^2 - w^2 = 10$, $z - w = 1 - i$ **b** $z^2 + w^2 = 5 - 2i$, $z + iw = 2 + 5i$ **c** $z^2 - w^2 = -2 + 16i$, $z + iw = 4 + 4i$

2.4 Real parts, imaginary parts and complex conjugates

Two complex numbers are equal if and only if their real parts are equal and their imaginary parts are equal.

Complex numbers do *not* behave like real numbers.
$2 + 5 = 3 + 4$ but $2 + 5i \neq 3 + 4i$

If $z = a + bi$ and $w = c + di$ with $a, b\ c, d \in \mathbb{R}$
then $z = w \Leftrightarrow a = c$ and $b = d$.

$\Leftrightarrow$ means 'implies and is implied by'.

EXAMPLE 1

Given that $a - 3i = 7 + bi$ where $a, b \in \mathbb{R}$, find the value of a and the value of b.

Equate the real parts of each number:

$\text{Re}(a - 3i) = a, \quad \text{Re}(7 + bi) = 7$

Since $a - 3i = 7 + bi$, the real parts must be equal.
Hence $\quad a = 7$

Equate the imaginary parts of each number:

$\text{Im}(a - 3i) = -3, \quad \text{Im}(7 + bi) = b$

Hence $\quad b = -3$

You can solve an equation in z by replacing z with the expression $a + bi$, where $a, b \in \mathbb{R}$.

A general complex number z has the form $z = a + bi$ for $a, b \in \mathbb{R}$.

EXAMPLE 2

Solve the equation $z^2 = 3 + 4i$

Let $z = a + bi$ then $z^2 = (a + ib)^2 = (a^2 - b^2) + 2abi$

$(a + ib)^2 = a^2 + 2abi + i^2b^2 = (a^2 - b^2) + 2abi$ is a useful expansion: You should learn it.

Replace z^2 with $(a^2 - b^2) + 2abi$ in the original equation:

$(a^2 - b^2) + 2abi = 3 + 4i$

Equate real parts: $\quad a^2 - b^2 = 3 \quad (1)$

Equate imaginary parts: $\quad 2ab = 4, \quad ab = 2 \quad (2)$

Make b the subject of (2): $\quad b = \frac{2}{a}$

Substitute $b = \frac{2}{a}$ into (1): $\quad a^2 - \frac{4}{a^2} = 3$

This is a quadratic equation in a^2.

Multiply through by a^2: $\quad a^4 - 3a^2 - 4 = 0$

$(a^2 + 1)(a^2 - 4) = 0$

so $a^2 = -1$ or $a^2 = 4$

$a = \pm 2$

$a^2 = -1$ has no solution since a is a *real* number.

Using $b = \frac{2}{a}$

When $a = 2$, $b = 1$ and so $z = 2 + i$
When $a = -2$, $b = -1$ giving $z = -2 - i$

Hence the solution is $z = 2 + i$ or $z = -2 - i$

You can write the solution as $z = \pm(2 + i)$. These numbers are the square roots of $3 + 4i$.

For any complex number $z = a + bi$, where $a, b \in \mathbb{R}$, the **complex conjugate** of z is $z^* = a - bi$

The conjugate of z is found by multiplying its imaginary part by (−1).

EXAMPLE 3

If $z = 1 + 3i$ and $w = -6 - 7i$ find

a z^* **b** w^* **c** $(z - w)^*$

a If $z = 1 + 3i$, $z^* = 1 - 3i$

b If $w = -6 - 7i$, $w^* = -6 + 7i$

c Since $z - w = (1 + 3i) - (-6 - 7i)$
$= 7 + 10i$

$(z - w)^* = 7 - 10i$

The real part is the same in a complex number, z, and its conjugate, z^*.
$\text{Re}(z^*) = \text{Re}(z)$
$\text{Im}(z^*) = -\text{Im}(z)$

You can solve an equation involving z^* by replacing z with $a + bi$, where $a, b \in \mathbb{R}$.

EXAMPLE 4

Solve the equation $4z - 2z^* = 12 + 3i$

If $z = a + bi$ then $z^* = a - bi$

Substitute $z = a + bi$ and $z^* = a - bi$ into the equation:

$$4z - 2z^* = 12 + 3i$$
$$4(a + bi) - 2(a - bi) = 12 + 3i$$

Collect real and imaginary parts:

$(4a - 2a) + (4b + 2b)i = 12 + 3i$

Simplify: $2a + 6bi = 12 + 3i$

Equate real and imaginary parts: $a = 6, b = \frac{1}{2}$

Hence the solution, $z = a + bi$, is given by $z = 6 + \frac{1}{2}i$

If $2a = 6$, $a = 3$

$6b = 3$ so $b = \frac{1}{2}$

For any complex number $z = a + bi$, where $a, b \in \mathbb{R}$

$z + z^* = 2a = 2\text{Re}(z)$ $z - z^* = 2bi = 2\text{Im}(z)i$ $zz^* = a^2 + b^2$

The notation zz^* means $z \times (z^*)$ **not** $(z \times z)^*$.

FP1

Exercise 2.4

Where appropriate, give answers in the form $a + ib$ where $a, b \in \mathbb{R}$.

1 Find the real numbers a and b for which

a $a + 2i = 7 + bi$ b $4 + bi = a - 2i$

c $3 - bi = a + 8i$ d $a - 5i = 3 - bi$

e $a + bi = 2(1 + 2i)$ f $a + bi = -2$

2 Find all real numbers a and b for which

a $4a + abi = 12 - 6i$ b $a + abi = 3(2 + i)$

c $a^2 + bi = a(4 - i)$ d $a(b + ai) = 3i$

e $a^2 + 2abi = 1 - 4i$ f $(a + ib)^2 = -12 + 4bi$

3 Given that $p(p + 5qi) = 4 - q^2 i$, where $p, q \in \mathbb{R}, q > 0$,

a show that $p = -2$, justifying your answer

b find the value of q.

FP1

4 Solve these equations.

a $z^2 = 15 + 8i$ b $z^2 = -21 + 20i$

c $z^2 - 12 + 16i = 0$ d $z^2 = 8i$

e $2z^2 - 4 + 3i = 0$ f $(2z + 1)(1 - 2z) = 8(2 - i)$

5 Write down the conjugate of each of these complex numbers.

a $3 + 7i$ b $-2 + 5i$

c $3 - 9i$ d $z^2 = 8i$

e $-(4 - 5i)$ f $\frac{3}{4}i$

6 Given that $z = 2 + 4i$, evaluate these expressions.

a $(2z)^*$ b $z - z^*$

c $2z + 3z^*$ d iz^*

e $i(z + iz^*)$

7 Solve these equations.

a $z + 3z^* = 4 - 6i$ b $2z + 5z^* = 21 + 6i$

c $3z - iz^* = 17 - 11i$ d $iz + 2z^* = 3(3 + 2i)$

e $z(2 + i) - z^* = 12i$ f $i(3z^* + iz) = z + 1$

8 Solve these pairs of simultaneous equations.

a $z + w^* = 8 + 7i$
$z^* - w = 4 - i$

b $z + iw = 1 + 4i$
$z^* + iw^* = 7 + 6i$

9 For the complex number $z = a + ib$, where $a, b \in \mathbb{R}$, $b \neq 0$
prove that $\frac{2bz}{z^* - z} = iz$

10 Prove these results for *any* complex number $z = a + ib$, where $a, b \in \mathbb{R}$.

a $z + z^* = 2a$

b $z - z^* = 2bi$

c $z(z^*) = a^2 + b^2$

d $(z^2)^* = (z^*)^2$

e $iz^* + (iz)^* = 0$

f $\left(\frac{1}{z}\right)^* = \frac{1}{z^*}$

11 If $z = a + bi$ and $w = b + ai$, where $a, b \in \mathbb{R}$ and $a + b \neq 0$

a simplify $\frac{z}{w^*}$

b prove that $\text{Re}\left(\frac{z}{z + w}\right) = \frac{1}{2}$

12 Given that $z = a + bi$, where $a, b \in \mathbb{R}$ are non-zero and $z^2 = c + di$ for $c, d \in \mathbb{R}$

a prove that $a^2 = \frac{1}{2}\left(c + \sqrt{c^2 + d^2}\right)$ and find a similar expression for b^2

b hence show that $(a^2 + b^2)^2 = c^2 + d^2$

c Apply the result of part **b** to $z = 5 + 12i$ to express 13^4 as the sum of two non-zero square numbers.

FP1

2.5 Representing complex numbers geometrically

You can represent a complex number geometrically using an **Argand diagram.**

The Argand diagram is a plane with a **real** horizontal axis and an **imaginary** vertical axis, which divide the plane into four quadrants.

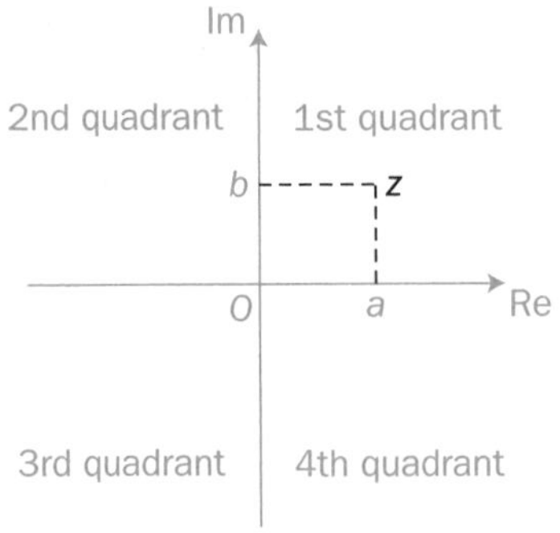

You can place a complex number $z = a + ib$ on the Argand diagram by plotting its real and imaginary parts on the relevant axes. In the diagram Re denotes the real axis and Im denotes the imaginary axis.

EXAMPLE 1

Represent the complex number $z = 2 + 3i$ on an Argand diagram.

Identify the real and imaginary parts of z:

$\text{Re}(z) = 2, \text{Im}(z) = 3$

Draw the diagram with real horizontal axis and imaginary vertical axis.

Plot a point with real coordinate 2 and imaginary coordinate 3.

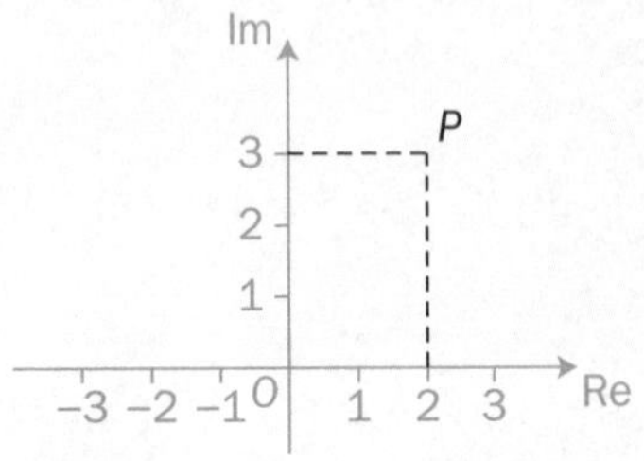

Point P on the Argand diagram represents the complex number $z = 2 + 3i$

You can refer to a complex number as a point in an Argand diagram.

EXAMPLE 2

Write down the complex numbers z and w represented by the points P and Q respectively on the Argand diagram.

You may assume integer values for each real and imaginary part.

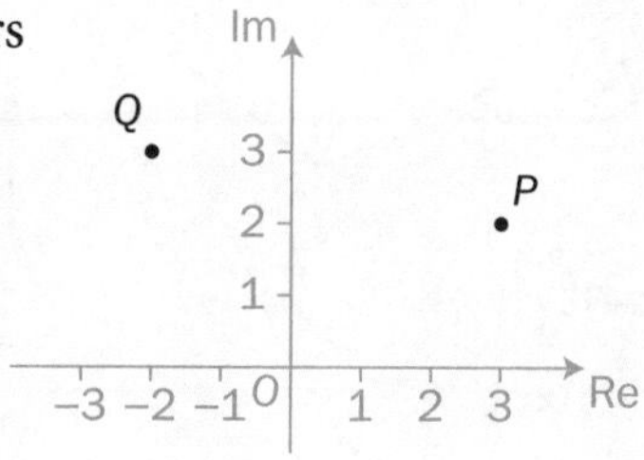

P has real coordinate 3 and imaginary coordinate 2 and so z has real part 3 and imaginary part 2.

Hence $z = 3 + 2i$

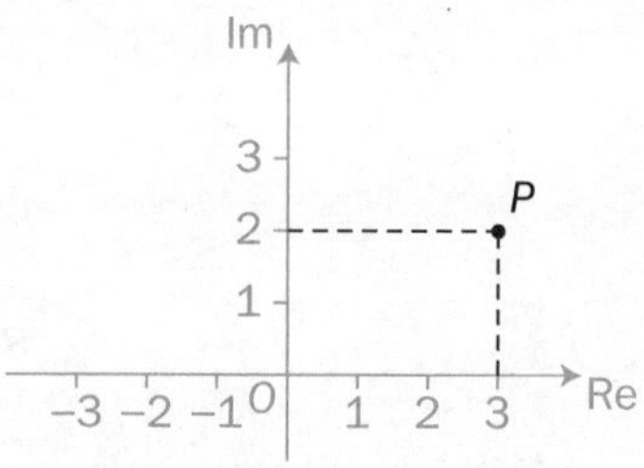

Similarly $w = -2 + 3i$

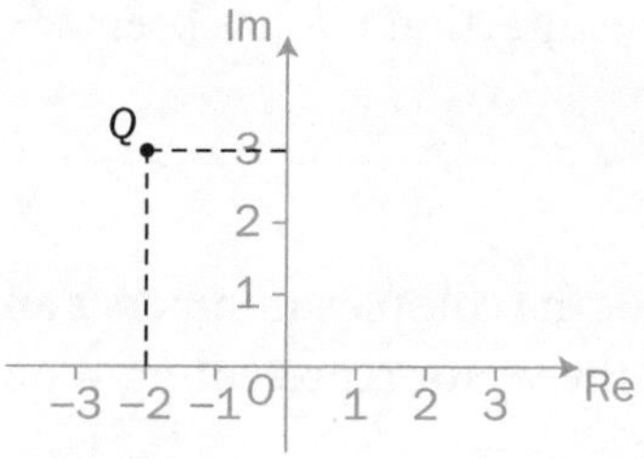

The point O on the Argand diagram where the axes cross represents the zero complex number $0 + 0i$.
The real and imaginary parts of this number are both zero.

You can represent a complex number z by drawing the vector from O to z.

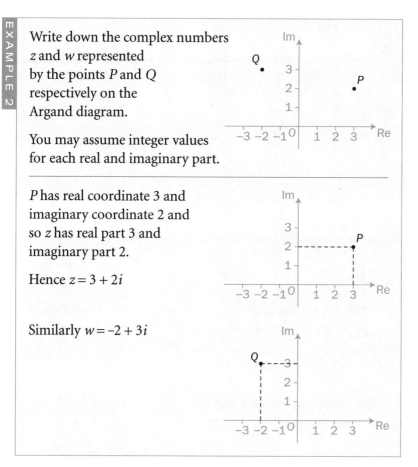

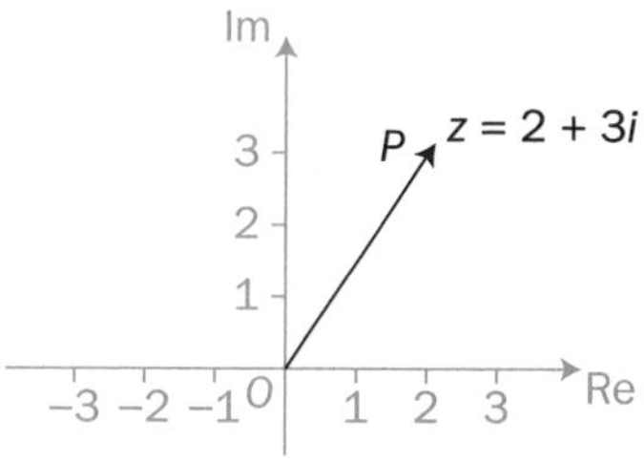

e.g. The vector $\overrightarrow{OP}$ represents the complex number $z = 2 + 3i$

Multiplying a complex number by a real number has a stretching effect on its geometric representation.

FP1

EXAMPLE 3

The points P, Q and R on the Argand diagram represent the complex numbers u, v and w respectively.
On a copy of the diagram, illustrate the complex numbers

a $2u$ **b** $\frac{1}{2}v$ **c** $-2w$

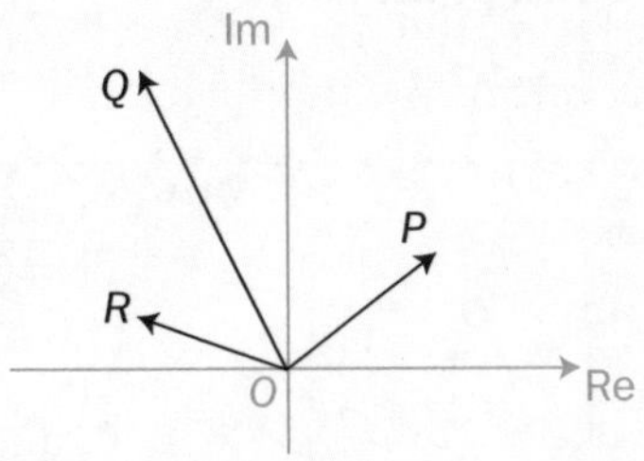

---

a $2u$ is represented by P', where OP' is a stretch, scale factor 2, of the vector OP.

b $\frac{1}{2}v$ is represented by Q', where OQ' is a stretch, scale factor $\frac{1}{2}$, of the vector OQ.

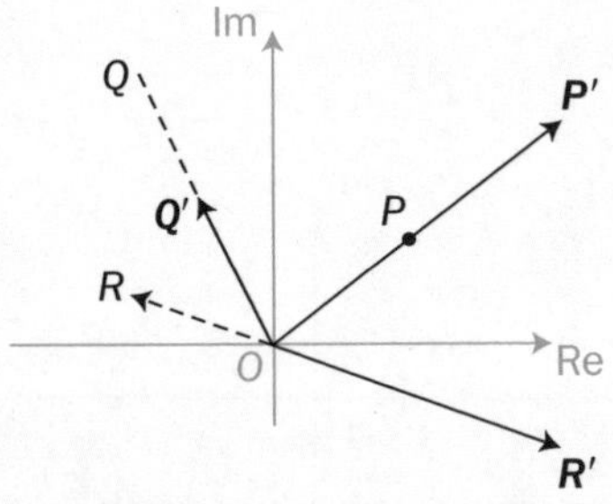

c $-2w$ is represented by R', where OR' is a stretch, scale factor 2, of the vector RO (the direction OR has been reversed).

Multiplying z by a negative number reverses the direction of the vector representing z.

FP1

You can represent the sum $z + w$, of the complex numbers z and w, on an Argand diagram by joining the vector representing w on to the end of the vector representing z.

The diagram shows vectors representing $z = 2 + 3i$, $w = 3 + i$ and their sum $z + w = 5 + 4i$.

The dotted line, which is a translation of the vector representing w, has been joined on to the end of the vector representing z. The result represents the sum $z + w$.

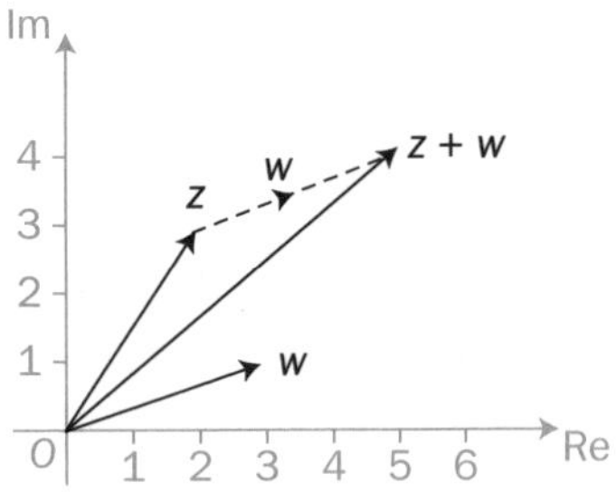

Since $z - w$ can be expressed as $z + (-w)$, you can represent $z - w$ geometrically by adding the vector representing $(-w)$ on to the vector representing z.

In the diagram, z and w are any pair of complex numbers. The dotted vector, which represents $-w$, has been joined on to the end of the vector representing z. The result is a representation of $z - w$.

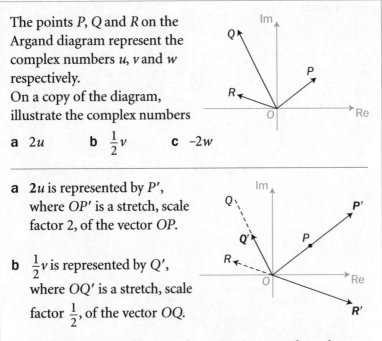

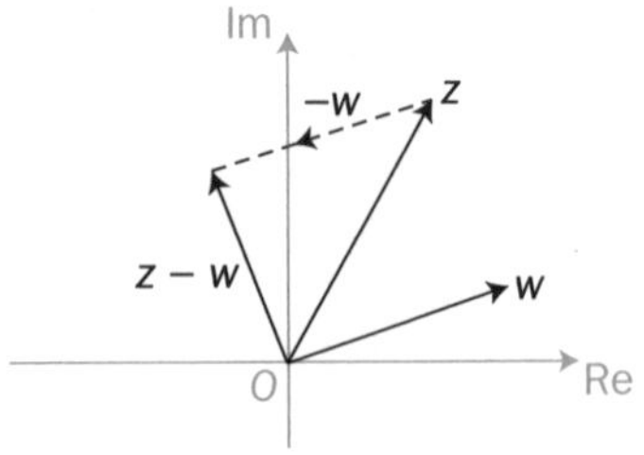

Reversing the direction of the vector representing w shows $-w$.

For the complex numbers z and w, the sum $z + w$ is constructed by joining the vector representing one of the numbers on to the vector representing the other.

The difference $z - w$ is constructed by joining the vector representing $-w$ on to the vector representing z.

Alternatively, to represent $z - w$, you can draw a line from w to z and then translate this line to the point O.

The alternative construction for $z - w$ is useful for solving geometrical problems.

EXAMPLE 4

Points P and Q in the Argand diagram represent the complex numbers z and w respectively, where $OP = 6$ and $OQ = 10$.

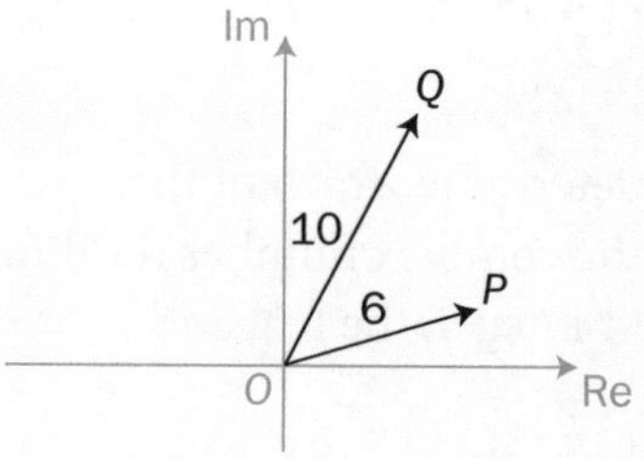

a Determine the quadrant to which $w - z$ belongs.

b Given that $OR = 8$, where R represents $w - z$, prove that triangle OPQ is right-angled.

---

a For $w - z$, draw a vector from P to Q and then translate that vector to the point O:

Using the alternative method of construction.

'$w - z$' means 'go from z to w'.

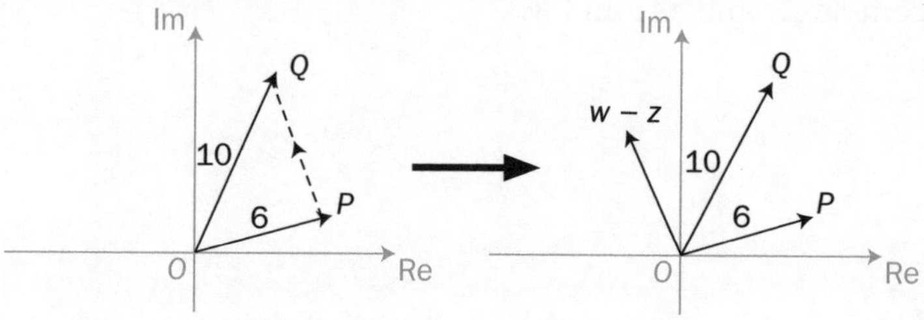

Hence $w - z$ lies in the second quadrant.

b The Argand diagram shows point R, which represents $w - z$. From the way in which $w - z$ was constructed, you can see that the length of OR equals the length of PQ.

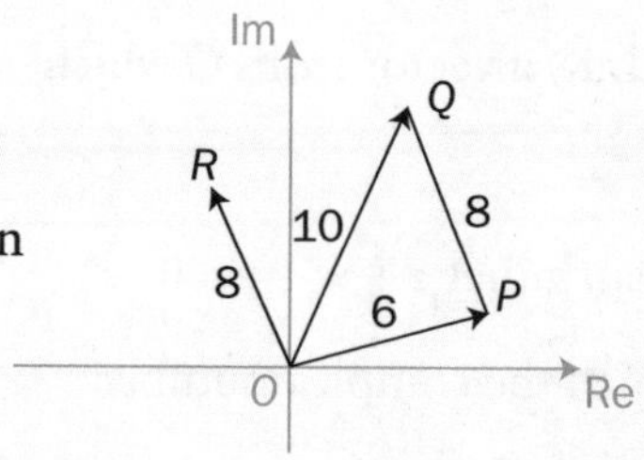

Hence $PQ = 8$ and so the side lengths 6, 8, and 10 of triangle OPQ obey Pythagoras' Theorem. This proves triangle OPQ is right-angled.

$6^2 + 8^2 = 36 + 64$
$= 100 = 10^2$

FP1

Exercise 2.5

1 The Argand diagram shows the complex numbers p, u, v, w and z.

You may assume that the real and imaginary parts of each number are integers.

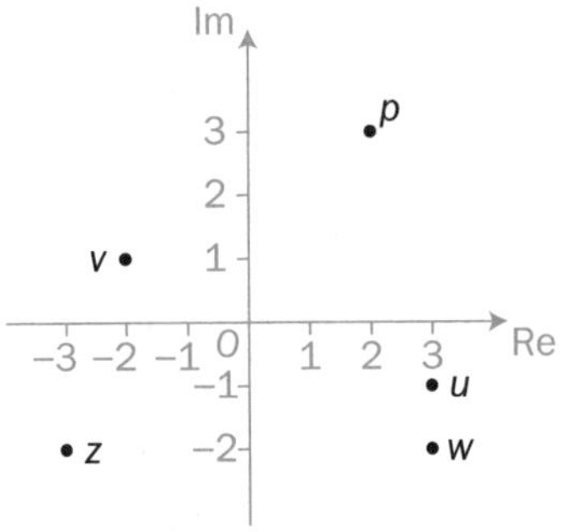

a Write down each number in the form $a + bi$ for integers a and b.

b Without using a calculator, determine which of the numbers on the diagram is the complex number resulting from rounding, to the nearest integer, the real and imaginary parts of $\pi - i\sqrt{3}$.

2 Show these complex numbers on the same Argand diagram.

a $z = -3 + 2i$ b $w = 1 - 2i$

c $u = -2 - 3i$ d $v = 2.5 + 3i$

e $p = -1$ f $q = -2.5i$

3 The Argand diagram shows the complex numbers z and w.

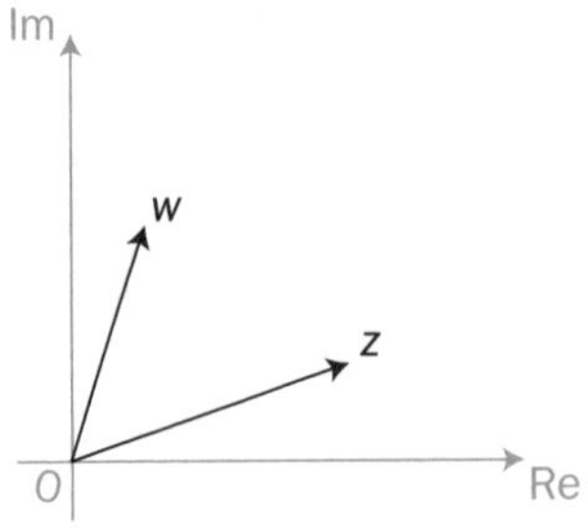

a On a copy of the diagram, draw a vector from O which represents

i $z + w$

ii the complex number v such that $z + w + v = 0$

b Identify the quadrant in which the complex number $w + v$ lies.

FP1

4 The Argand diagram shows points A and B, which represent the complex numbers z and w respectively.

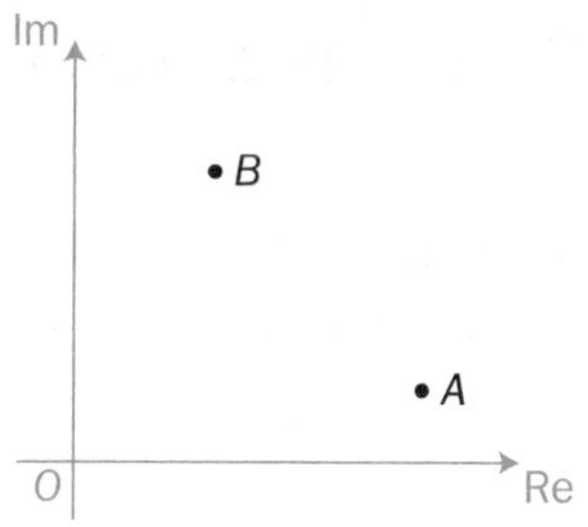

a On a copy of the diagram, indicate
  i the point C representing $z + w$
  ii the point D representing $2z$
  iii the point E representing $2z + w$.

b Completely describe the quadrilateral $OCED$.

5 For the complex numbers z and w, the vectors OP and OQ on the Argand diagram represent $z + w$ and $z - w$ respectively. You may assume that P and Q are correctly positioned relative to the value of the constant k marked on each axis. Use an Argand diagram to demonstrate that

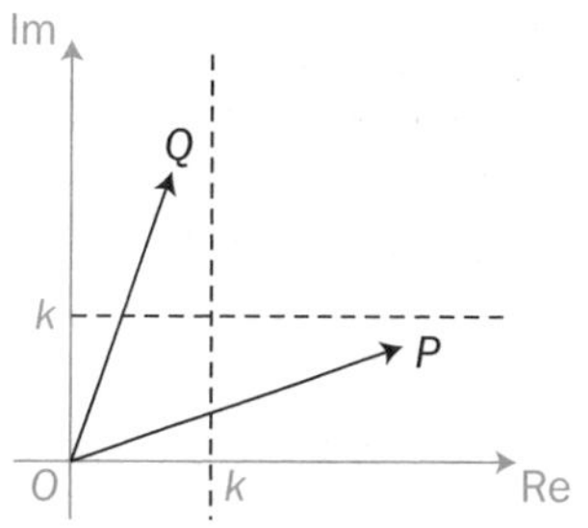

a z lies in the first quadrant

b w lies in the fourth quadrant.

6 In the Argand diagram, the complex numbers z and w, in the first quadrant, are represented by the points P and Q respectively.
P and Q lie on a common circle, centre O, part of which is shown in the diagram.

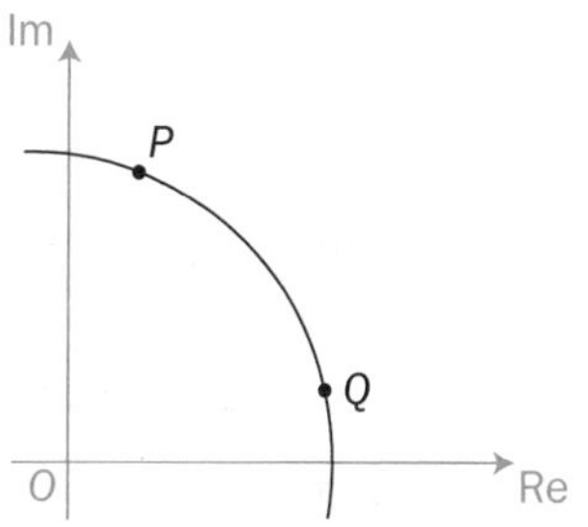

a On a copy of the diagram, draw a vector
  i OA to represent $z + w$
  ii OB to represent $w - z$.

b Completely describe the quadrilateral $OPAQ$.

c Given that point B lies on this circle, show that angle $POQ = 60°$.

7 The Argand diagram shows vectors OP and OQ in the first quadrant, representing the non-zero complex numbers z and w respectively. You may assume that the position of Q is correct relative to P.

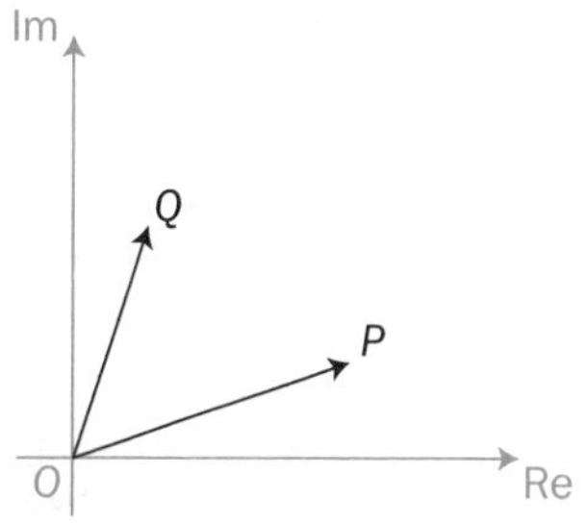

Points A and B, not shown on the diagram, represent $z + w$ and $z - w$ respectively.

Given that the lengths OA and OB are equal, show that z is real and that w is purely imaginary.

FP1

2.6 The modulus and argument of a complex number

You can define the modulus and argument of a complex number $z = a + bi$ using the vector that represents z on an Argand diagram.

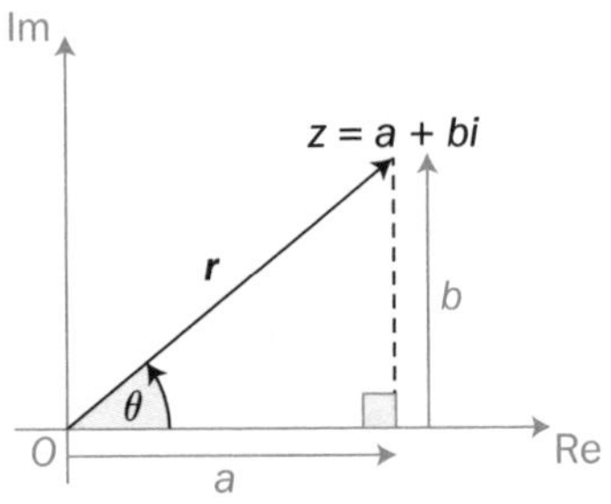

The modulus of a complex number z, written $|z|$, is the length r of the vector which represents z.

An angle, θ, measured from the positive real axis to the vector representing z is called an argument of z, written as $\arg z$.

If $z = a + bi$ then $|z| = \sqrt{a^2 + b^2}$

You can show this using Pythagoras' Theorem.

If $\arg z = \theta$ then $\tan \theta = \frac{b}{a}$

You can show this using $\tan \theta = \frac{\text{opposite}}{\text{adjacent}}$

EXAMPLE 1

Find the modulus and an argument of $z = 2 + 2i$.

Illustrate z on an Argand diagram:

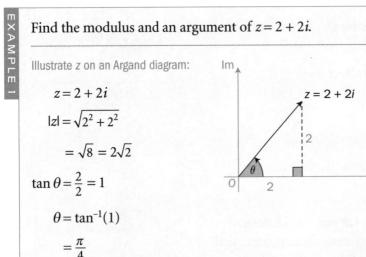

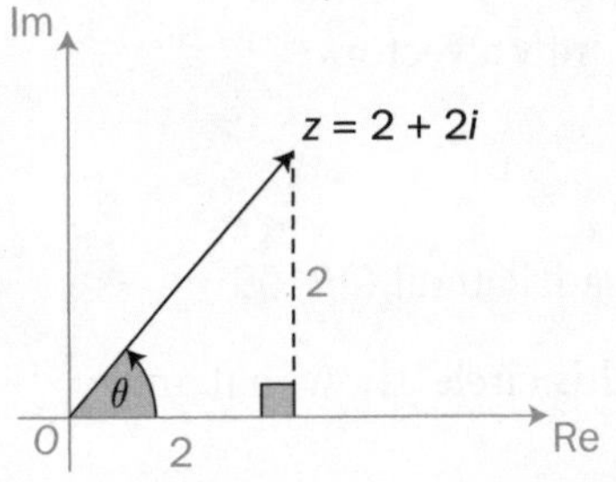

$z = 2 + 2i$

$|z| = \sqrt{2^2 + 2^2}$

$= \sqrt{8} = 2\sqrt{2}$

$\tan \theta = \frac{2}{2} = 1$

$\theta = \tan^{-1}(1)$

$= \frac{\pi}{4}$

Hence $|z| = 2$, $\arg z = \frac{1}{4}\pi$

An argument of a complex number is negative if it is measured in a clockwise direction.

The diagram shows a complex number z in the second quadrant. θ_1 and θ_2 are two possible arguments of z where $\theta_1 > 0$ and $\theta_2 < 0$.

You can see from this that a complex number has more than one argument.

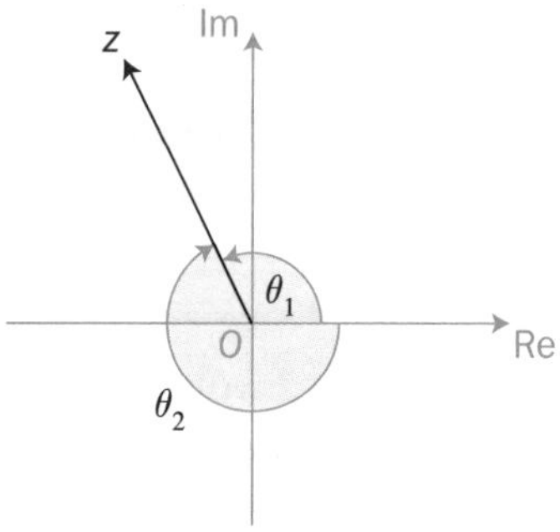

An argument θ of z is **principal** if $-\pi < \theta \leqslant \pi$.
Every complex number has exactly one **principal argument**.
Unless told otherwise you should assume that $\arg z$ refers to the principal argument of z.

You can evaluate $\arg z$, where $z = a + bi$, (for $a \neq 0$) by calculating $\tan^{-1}\left(\frac{b}{a}\right)$ and then adjusting the answer, if necessary, according to the quadrant in which z lies.

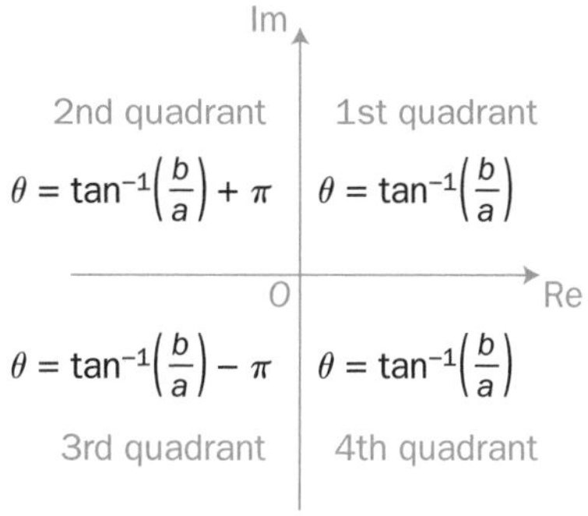

FP1

EXAMPLE 2

Find, in exact form, $\arg(z)$ when $z = -1 + i\sqrt{3}$.

Compare $z = -1 + \sqrt{3}i$ with $z = a + bi$:

$a = -1$ and $b = \sqrt{3}$

$$\tan^{-1}\left(\frac{b}{a}\right) = \tan^{-1}\left(\frac{\sqrt{3}}{-1}\right) = -\frac{1}{3}\pi$$

Use the result $\tan^{-1}(-\theta) \equiv -\tan^{-1}\theta$

Draw an Argand diagram to help you determine the quadrant in which z lies:

Since the complex number $-1 + i\sqrt{3}$ lies in the second quadrant,

$$\arg z = \tan^{-1}\left(\frac{b}{a}\right) + \pi = -\frac{1}{3}\pi + \pi$$

$$= \frac{2}{3}\pi$$

The principal argument of z is $\arg z = \frac{2\pi}{3}$

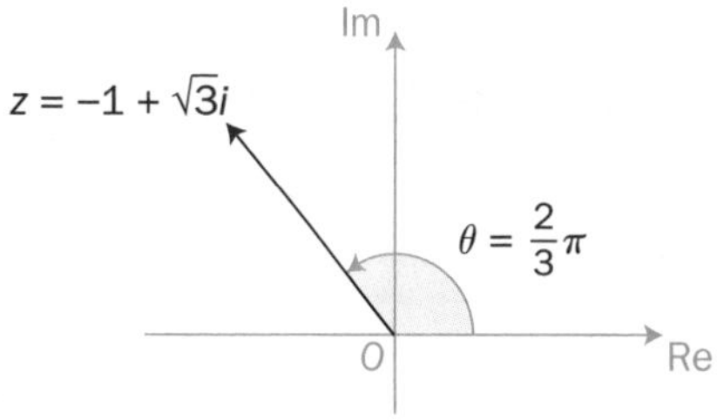

If $z = a + bi$ has zero real part, then $a = 0$ giving $z = bi$.

In this case, $\arg z = \pm\frac{\pi}{2}$ depending on the sign of b.

The argument of the zero complex number, $z = 0 + 0i$, is undefined.

You can solve geometrical problems involving complex numbers by using properties of the modulus and argument.

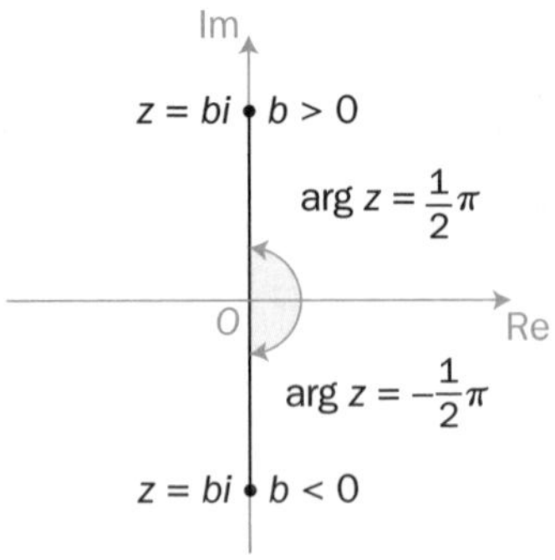

EXAMPLE 3

In the Argand diagram, point P represents $z = -3 + 2i$ and point Q represents $w = 2 + 4i$

a Find the exact distance PQ.

b Find angle QOP. Give your answer in radians, correct to one decimal place.

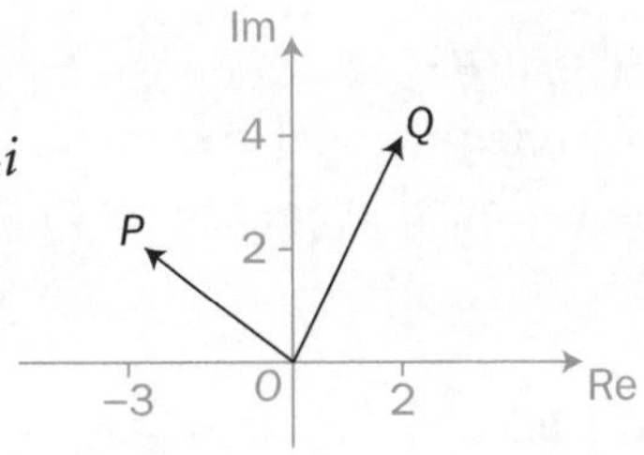

---

a Draw an Argand diagram to help you visualise the problem:

Refer to Section 2.5.

PQ is equal to the modulus of the complex number $(w - z)$.

$|w - z|$ = length of vector joining w and z

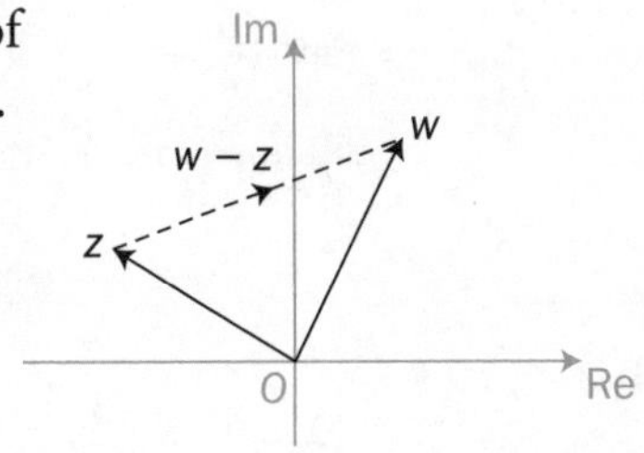

$$w - z = (2 + 4i) - (-3 + 2i)$$
$$= 5 + 2i$$

$$PQ = |w - z|$$
$$= |5 + 2i|$$
$$= \sqrt{29}$$

$\sqrt{5^2 + 2^2} = \sqrt{29}$

b Use a diagram to see which angle you need to find:

Put your calculator in radian mode.

Angle $QOP = \alpha = \arg z - \arg w$

Calculate arg z and arg w:

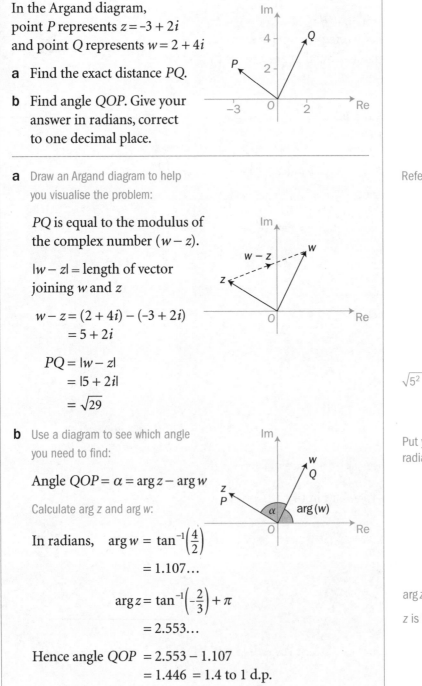

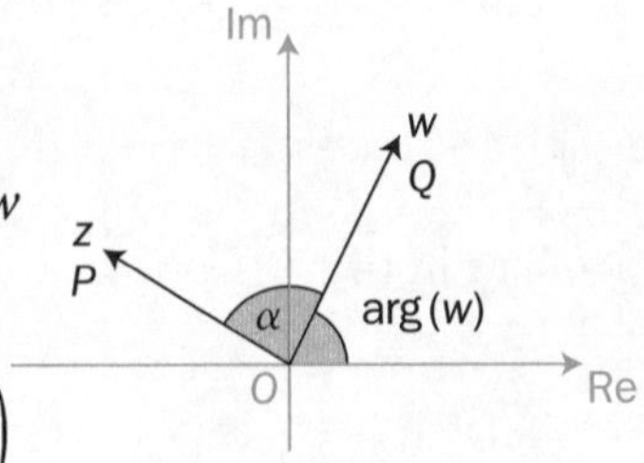

In radians, $\arg w = \tan^{-1}\left(\frac{4}{2}\right)$

$$= 1.107\ldots$$

$$\arg z = \tan^{-1}\left(-\frac{2}{3}\right) + \pi$$

$$= 2.553\ldots$$

$\arg z = \tan^{-1}\left(-\frac{2}{3}\right) + \pi$ since z is in the second quadrant.

Hence angle $QOP = 2.553 - 1.107$

$$= 1.446 = 1.4 \text{ to 1 d.p.}$$

FP1

Exercise 2.6

All arguments in these questions are principal and are in radians.

1 Find the modulus and argument of each of these complex numbers.
Give your answers to one decimal place.

a $z = 3 + 7i$ b $z = -4 + 5i$

c $z = 6 - 2i$ d $z = -1 - 3i$

e $z = \frac{3}{4} - \frac{2}{3}i$ f $z = -\sqrt{2} + 4.2i$

2 Find the modulus and argument of each complex number.
Give your answers for $\arg z$ correct to one decimal place.

a $z = 3 + 4i$ b $z = -5 + 12i$

c $z = 10 - 10.5i$ d $z = -4.5 - 6i$

e $z = \frac{1}{9}(2 + \sqrt{5}i)$ f $z = -\sqrt{3} + \sqrt{6}i$

3 Find the modulus and argument of each of these complex numbers.
Give $|z|$ in simplified surd form (where appropriate) and $\arg z$ in terms of π.

a $z = 3 + \sqrt{3}i$ b $z = -2 + 2i$

c $z = -\sqrt{2} - i\sqrt{6}$ d $z = \frac{\sqrt{3}}{2} - \frac{3}{2}i$

e $z = \sqrt{5} + i\sqrt{15}$ f $z = -\frac{1}{\sqrt{2}} + \frac{\sqrt{2}}{2}i$

g $z = \sqrt{2}$ h $z = -i\sqrt{18}$

4 For the complex numbers $z = 3 + i$ and $w = 1 + 4i$, calculate these quantities.
Give your answers correct to two decimal places where appropriate.

a $|z + w|$ b $\arg(z - w)$

c $|iz + w|$ d $\arg(z + iw)$

e $|zw - 1|$ f $\arg(zw - i)$

FP1

5 The complex number $z = a + 3i$, where $a < 0$ is a real number, has modulus 5.

a Find the value of a.

b Find the argument of z.
Give your answer to one decimal place.

6 Given that $z = 2 + bi$, where $b \in \mathbb{R}$, has argument $-\frac{1}{3}\pi$

a find the exact value of b

b calculate $|z|$.

7 Find the modulus and argument of each of these complex numbers. Give your answers in exact form.

a $z = \dfrac{1}{1 - i\sqrt{3}}$

b $z = \dfrac{3}{2 + 2i}$

c $z = \dfrac{1 - i}{i}$

d $z = \dfrac{\sqrt{2}}{1 + i}$

e $z = \dfrac{i}{1 - i\sqrt{3}}$

f $z = \dfrac{3}{i(\sqrt{3} + i)}$

8 The complex number z satisfies the equation $\dfrac{z + \alpha}{z - i\alpha} = 2$, where $\alpha \in \mathbb{R}$ is negative.

a Show that $z = \alpha(1 + 2i)$

b Find $\arg z$, giving your answer to two decimal places.

c Express $|z|$ in terms of α.

9 A and B on the Argand diagram represent the complex numbers $z = \sqrt{2} + \sqrt{6}i$ and $w = 3 - i\sqrt{3}$ respectively.

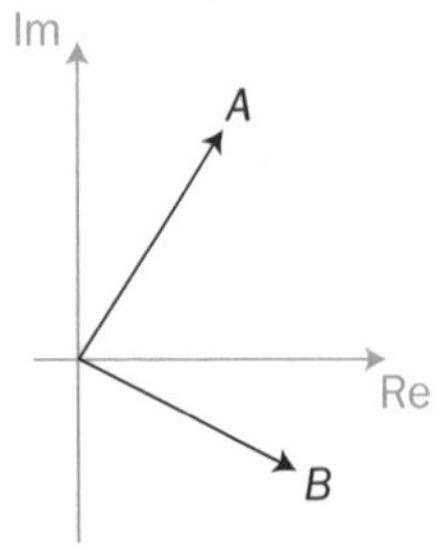

a Show that triangle OAB is right-angled.

b Hence show that the area of triangle OAB is $2\sqrt{6}$.

FP1

10 For each of the following, calculate the distance AB.
Give your answers in surd form.

a

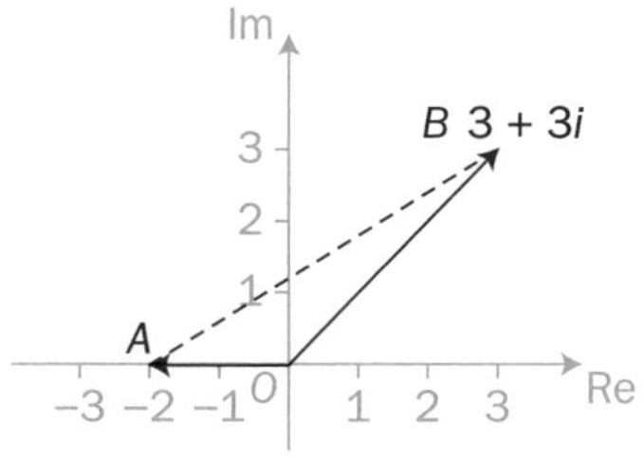

b

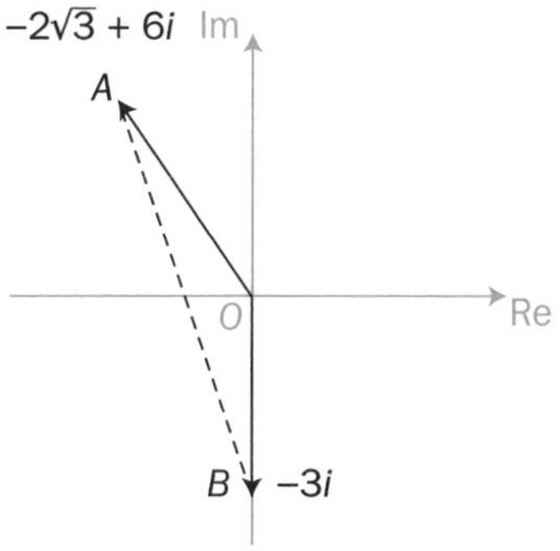

11 Points P and Q on the Argand diagram represent the complex numbers $z = 2(1 + i\sqrt{3})$ and $w = -\sqrt{3} + i$ respectively.

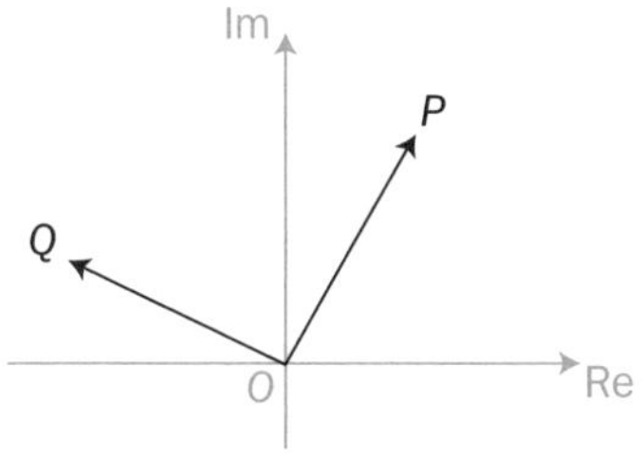

a Find, in terms of π, $\arg z$ and $\arg w$.

b Hence show that angle $POQ = \frac{1}{2}\pi$

c Find the area of triangle OPQ.

d Use a *geometrical* approach to show that, for this particular pair of numbers,
$|z + w|^2 = |z|^2 + |w|^2$

12 On the Argand diagram, points P and Q represent the complex numbers $z = \sqrt{3} + i$ and $w = -3 + \sqrt{3}\,i$ respectively.

The obtuse angle $\theta = POQ$ has been marked on the diagram.

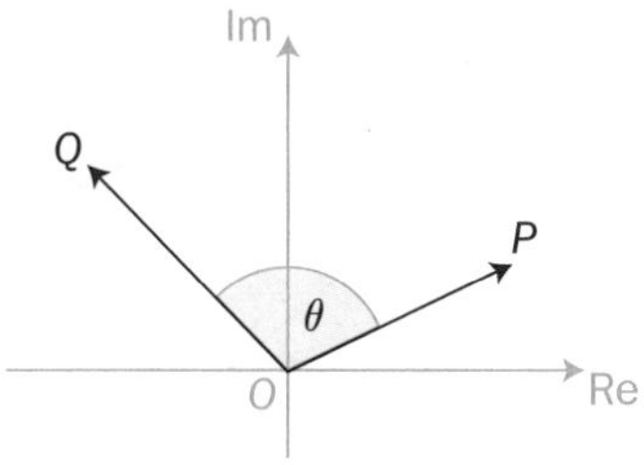

a Show that $\theta = \frac{2}{3}\pi$

b Find the acute angle OPQ, giving your answer to two decimal places.

2.7 The modulus-argument form for complex numbers

You can express a complex number in modulus-argument form.

In the Argand diagram, the complex number $z = a + ib$ has modulus r and argument θ.

Refer to Section 2.6.

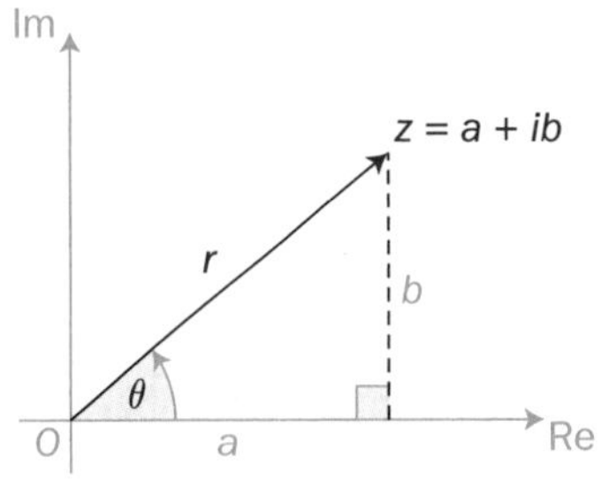

The complex number $z = a + ib$ is in **cartesian form** since you position z on the Argand diagram in the same way that you plot a point with cartesian coordinates (a, b) on a pair of axes.

You can see that $\cos\theta = \frac{a}{r}$, so $a = r\cos\theta$

and $\sin\theta = \frac{b}{r}$, so $b = r\sin\theta$

Hence $z = a + ib$ — z is in cartesian form.

$= r\cos\theta + i(r\sin\theta)$

$\Rightarrow \quad z = r(\cos\theta + i\sin\theta)$ — z is in modulus-argument form.

> If z has modulus r and argument θ then in modulus-argument form, $z = r(\cos\theta + i\sin\theta)$

You can convert a complex number from cartesian form into modulus-argument form and vice versa.

EXAMPLE 1

Express $z = 3 - 3i$ in modulus-argument form.

In modulus-argument form, $z = r(\cos\theta + i\sin\theta)$ where $r = |z|$ and $\theta = \arg z$

Drawing a diagram will help you to identify the quadrant in which z lies.

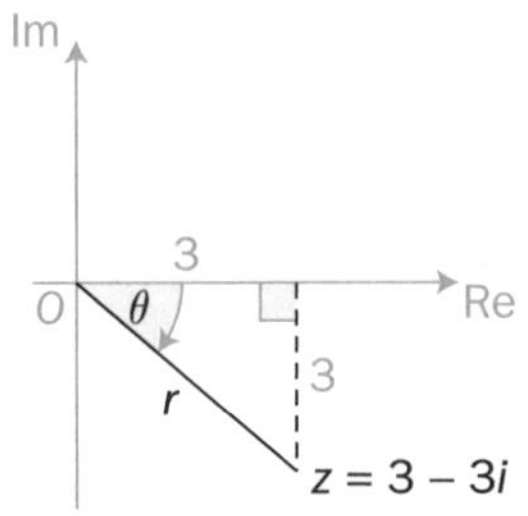

$z = 3 - 3i \Rightarrow r = |z| = \sqrt{3^2 + (-3)^2} = \sqrt{18} = 3\sqrt{2}$

$z = 3 - 3i$ lies in the fourth quadrant

$\Rightarrow \arg z = \tan^{-1}\left(-\frac{3}{3}\right)$

Use $\tan^{-1}(-\alpha) = -\tan^{-1}(\alpha)$: $\theta = \arg z = -\frac{1}{4}\pi$

Hence in modulus-argument form

$z = 3\sqrt{2}\left(\cos\left(-\frac{1}{4}\pi\right) + i\sin\left(-\frac{1}{4}\pi\right)\right)$

Leave $\cos\left(-\frac{1}{4}\pi\right)$ and $\sin\left(-\frac{1}{4}\pi\right)$ in the answer.

EXAMPLE 2

The complex number z has modulus 8 and argument $\frac{5}{6}\pi$.
Find z in cartesian form.

$r = 8, \quad \theta = \frac{5}{6}\pi$

Hence $z = r(\cos\theta + i\sin\theta)$

$= 8\left(\cos\left(\frac{5}{6}\pi\right) + i\sin\left(\frac{5}{6}\pi\right)\right)$

$= 8\left(-\frac{1}{2}\sqrt{3} + \frac{1}{2}i\right)$

$\Rightarrow \quad z = -4\sqrt{3} + 4i$

Refer to C2

$\cos\left(\frac{5}{6}\pi\right) = -\cos\left(\frac{1}{6}\pi\right) = -\frac{\sqrt{3}}{2}$

$\sin\left(\frac{5}{6}\pi\right) = \sin\left(\frac{1}{6}\pi\right) = \frac{1}{2}$

Draw an Argand diagram to check your answer is feasible.

If z and w are two complex numbers then the modulus of their product is

$|zw| = |z||w|$

and, provided that $w \neq 0$, the modulus of their quotient is

$\left|\frac{z}{w}\right| = \frac{|z|}{|w|}$

EXAMPLE 3

If $z = \sqrt{3} + i$ and $w = 2 - 2i$ calculate

a $|zw|$ **b** $|z^2|$ **c** $\left|\frac{w^2}{z^3}\right|$

First find the modulus of each given complex number:

$z = \sqrt{3} + i \Rightarrow |z| = \sqrt{3+1} = 4$

$w = 2 - 2i \Rightarrow |w| = \sqrt{4+4} = \sqrt{8}$

$\sqrt{8} = 2\sqrt{2}$

a $|zw| = |z||w|$

$= 4 \times 2\sqrt{2} = 8\sqrt{2}$

b $|z^2| = |z \times z|$

$= |z| \times |z|$

$= 4 \times 4 = 16$

This shows that $|z^2| = |z|^2$
In general, $|z^n| = |z|^n$
where n is a positive integer.

c $\left|\frac{w^2}{z^3}\right| = \frac{|w|^2}{|z|^3}$

$= \frac{8}{64} = \frac{1}{8}$

$|w^2| = |w|^2, \ |z^3| = |z|^3$

FP1

Exercise 2.7

1 Express these complex numbers in modulus-argument form.
In each case give the modulus in simplified surd form and the argument to two decimal places.

a $z = 1 + 3i$ b $z = 2 - 5i$

c $z = -\frac{1}{3} + \frac{1}{2}i$ d $z = -2 - 4i$

e $z = \sqrt{2} + 3i$ f $z = \frac{1}{\sqrt{2}} - \frac{1}{\sqrt{3}}i$

2 Express each number in cartesian form.
Give real and imaginary parts of each number to two decimal places.

a z, where $|z| = 3$ and $\arg z = 1.5^c$

The symbol ^{c} stands for radians.

b w, where $|w| = 6$ and $\arg w = -2.8^c$

c u, where $|u| = 1.5$ and $\arg u = \frac{7\pi}{12}$

Angles which involve π are assumed to be in radians.

d v, where $|v| = \sqrt{10}$ and $\arg v = -\frac{5\pi}{11}$

3 Copy and complete the following table.
Give exact answers where appropriate.

Modulus-argument form	Cartesian form
$z = 4\left(\cos\left(\frac{1}{3}\pi\right) + i\sin\left(\frac{1}{3}\pi\right)\right)$	$z =$
w	$w = -4 + 4i$
$p = \sqrt{6}\left(\cos\left(-\frac{1}{6}\pi\right) + i\sin\left(-\frac{1}{6}\pi\right)\right)$	$p =$
$q =$	$q = -\sqrt{3} - i$

4 $z = 1 + i$ and $w = 1 - i\sqrt{3}$

a Find the exact values of $|z|$ and $|w|$.

b Hence find the exact value of

i $|z^3w|$ ii $\left|\frac{z^5}{w^2}\right|$

c Show that $|z - w| = 1 + \sqrt{3}$ and hence find the exact value of $|z^2 - zw|$.

5 It is given that $z = 3 + \sqrt{3}i$ and $\left|\frac{z}{w}\right| = 2\sqrt{6}$ for w a particular complex number.

a Show that $|w| = \frac{1}{2}\sqrt{2}$

b Given further that $\arg w = -\frac{1}{4}\pi$, express w in

i modulus-argument form,
ii cartesian form.

6 $z = \sqrt{3}\left(\cos\left(\frac{1}{3}\pi\right) - i\sin\left(\frac{1}{3}\pi\right)\right)$

a Briefly explain why this expression is not the modulus-argument form for z.

b Express z in exact cartesian form.

c Hence, or otherwise, express z in modulus-argument form.

7 $z = \frac{1}{4}\left(\cos\left(\frac{2}{3}\pi\right) + i\sin\left(\frac{2}{3}\pi\right)\right)$ and $w = 4 + bi$ where $b < 0$.
It is given that $|zw^2| = 16$

a Show that $b = -4\sqrt{3}$ and hence express w in exact modulus-argument form.

b Illustrate, on a single Argand diagram, the complex numbers z and w.

c Hence, or otherwise, find the value of $|z + w|$.

8 $z = 1 + i$, $w = 2 + i$ and $u = 3 + i$

a Show that $|zw| = |u|$ and find the value of $\left|\frac{wu}{z}\right|$

b Given that $\arg(wu) = \arg w + \arg u$, find the exact value of $\tan^{-1}\left(\frac{1}{2}\right) + \tan^{-1}\left(\frac{1}{3}\right)$

c Illustrate, on a single Argand diagram, the points A and B representing the numbers z and w respectively.

d Find the area of triangle OAB and hence show that $\sin B\hat{O}A = \frac{1}{\sqrt{10}}$.

9 For $z = 1 + \sqrt{3}i$, use the information given to find, in exact cartesian form,

a w, given that $|zw| = \sqrt{32}$ and $\arg(w) = -\frac{1}{4}\pi$

b v, given that $\left|\frac{v^2}{z}\right| = 2$ and $\arg v = \frac{5}{6}\pi$

10 Given that $|z| = 3$ and $w = 1 + 2i$, find the exact value of

a $|zw - z|$

b $|z + zw|$

c $\left|\dfrac{w - i}{z^2}\right|$

d $|w^2 - 1|$

11 In the Argand diagram, points A and B represent the complex numbers $z = 1 + \sqrt{3}i$ and $w = 2 - 2i$ respectively.

Point C represents zw.

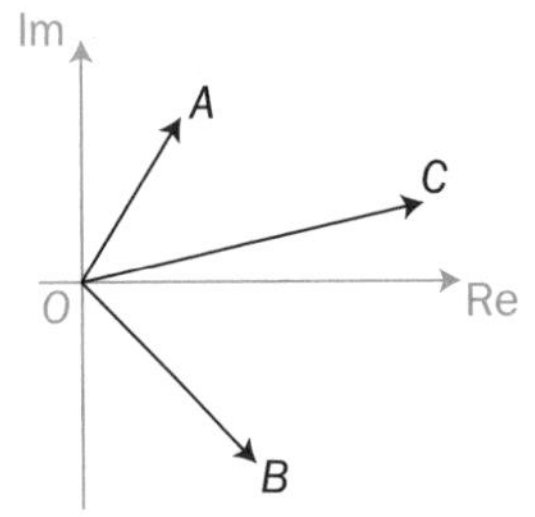

a Find the exact value of $|zw|$.

b Given that $\arg(zw) = \dfrac{1}{12}\pi$, find the exact area of quadrilateral $OACB$.

12 Points A and B on the Argand diagram represent the complex numbers $z = \sqrt{3} + i$ and $w = \dfrac{3\sqrt{2}}{2}(-1 + i)$ respectively.

Point C represents $\dfrac{w}{z}$.

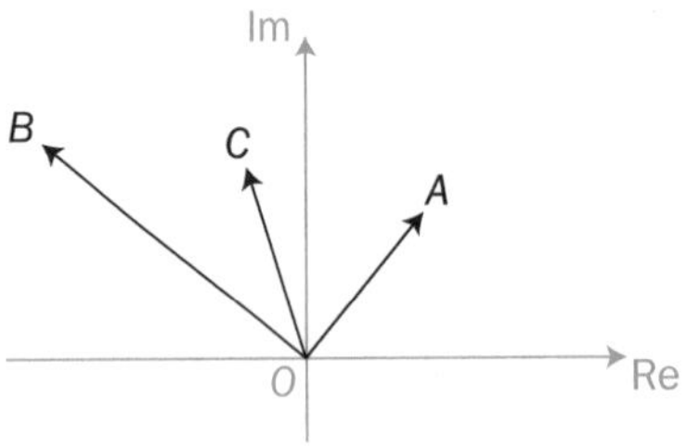

a Express z and w in modulus-argument form.

b Given that $\arg\left(\dfrac{w}{z}\right) = \arg w - \arg z$, show that angle $COB = \dfrac{1}{6}\pi$

c Hence

i find the area of triangle OBC

ii calculate the distance AC giving your answer correct to two decimal places.

13 In the Argand diagram, point A represents $z = 1 + i$ and point B represents zw for w, a particular complex number.

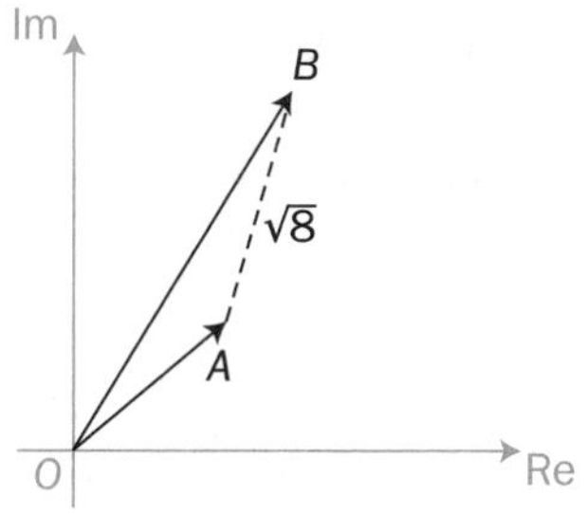

The line AB has length $\sqrt{8}$.

a State the value of $|zw - z|$ and hence find the value of $|w - 1|$.

b Given further that $\arg(w - 1) = \frac{1}{6}\pi$, find w in exact cartesian form.

14 In the Argand diagram, points A, B and C represent the complex numbers z, w and $\frac{w}{z}$ respectively.

$OA = 2$, $OB = 5$ and the angle between OB and OC is $\frac{1}{4}\pi$ radians.

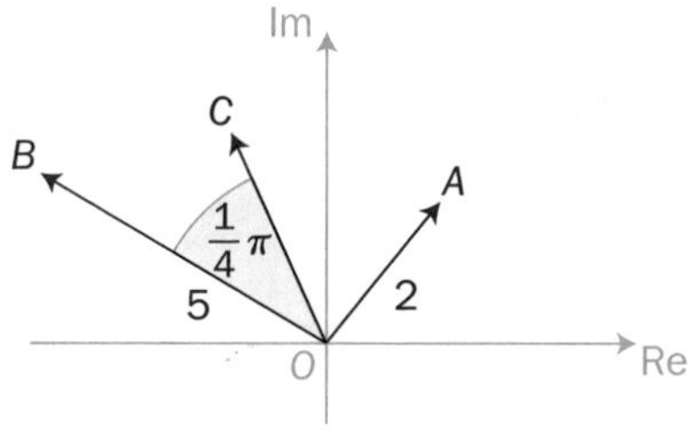

Given that $AB^2 = 39$ find, in modulus-argument form, the complex number w.
You may assume the result $\arg\left(\frac{w}{z}\right) = \arg w - \arg z$

FP1

2.8 Roots of polynomial equations

A polynomial in the complex number z is an expression involving non-negative integer powers of z.

e.g. $P(z) = 2z^4 - \pi z^2 + 1$ is a polynomial in z.

EXAMPLE 1

If $P(z) = 5z^2 - 2z + 1$, evaluate $P(1 + i)$.

$P(z) = 5z^2 - 2z + 1$

Substitute $z = 1 + i$:

$$\begin{aligned}P(1 + i) &= 5(1 + i)^2 - 2(1 + i) + 1\\ &= 5(2i) - 2(1 + i) + 1\\ &= 10i - 2 - 2i + 1\\ &= -1 + 8i\end{aligned}$$

Therefore $P(1 + i) = -1 + 8i$

You have to find the value of $5z^2 - 2z + 1$ when $z = 1 + i$

$(1 + i)^2 = (1 + i)(1 + i)$
$= 1 + 2i + i^2 = 2i$

Given the polynomial $P(z)$, any value α for which $P(\alpha) = 0$ is called a root of the equation $P(z) = 0$.

$P(\alpha) = 0$ means $P(\alpha) = 0 + 0i$, the zero complex number.

EXAMPLE 2

Verify that $1 + 2i$ is a root of the equation $z^2 - 2z + 5 = 0$

Define $P(z) = z^2 - 2z + 5$

Substitute $z = 1 + 2i$:

$$\begin{aligned}P(1 + 2i) &= (1 + 2i)^2 - 2(1 + 2i) + 5\\ &= (-3 + 4i) - 2(1 + 2i) + 5\\ &= 0 + 0i\end{aligned}$$

Since $P(1 + 2i) = 0$, the complex number $1 + 2i$ is a root of the equation $z^2 - 2z + 5 = 0$

If $1 + 2i$ is a root of the equation $z^2 - 2z + 5 = 0$, then $P(1 + 2i) = 0$

$(1 + 2i)^2 = 1 + 4i + 4i^2$
$= 1 + 4i - 4$
$= -3 + 4i$

Exercise 2.8

1 Given that $P(z) = z^2 + z + 1$, evaluate each expression. Give your answers in the form $a + bi$ for $a, b \in \mathbb{R}$. Use exact values for a and b.

a $P(i)$ **b** $P(2i)$

c $P\left(-\frac{1}{2}i\right)$ **d** $P(i\sqrt{2})$

FP1

2 Evaluate

a $P(3i)$ for $P(z) = 2z^2 + 3z - 1$

b $P(2 + 3i)$ for $P(z) = 3z^2 - 2z + 4$

c $P(2i)$ for $P(z) = z^3 + z^2 + z + 1$

d $P(-1 + 2i)$ for $P(z) = z^3 + 2z - 1$

3 Verify that each complex number, α, is a root of the given equation.

a $\alpha = 1 + i$, $z^2 - 2z + 2 = 0$

b $\alpha = 3 + 2i$, $z^2 - 6z + 13 = 0$

c $\alpha = 2 + i$, $z^3 - 5z^2 + 9z - 5 = 0$

d $\alpha = 1 - i$, $z^4 + 4 = 0$

4 a Given that $2 - i$ is a root of the equation $z^2 + az + (1 - a) = 0$, find the value of the real constant, a.

b Given that $3i$ is a root of the equation $P(z) = 0$, where $P(z) = z^3 - z^2 + az + b$, find the value of the real constants, a and b.

5 Given that $2 + 3i$ is a root of the equation $P(z) = 0$, where $P(z) = z^2 + az + b$, for $a, b \in \mathbb{R}$

a show that $a = -4$ and find the value of b

b find the value of the real number k for which $z = 2 + ki$ is another root of the equation $P(z) = 0$

6 Find the value of real constants, a and b, such that $a + i$ is a root of the equation $z^2 - 4z + b = 0$

7 It is given that $1 - i$ is a root of the equation $P(z) = 0$, where $P(z) = z^3 + kz + h$, for real constants h and k.

a Show that $k = -2$ and find the value of h.

b Show that $P(z + 1) = z^3 + 3z^2 + z + 3$ and hence write down a complex number which is a root of the equation $z^3 + 3z^2 + z + 3 = 0$

8 a Verify that $1 + i$ is a root of the equation $z^4 + z^2 - 2z + 6 = 0$

b Hence, by making the substitution $w = \frac{1}{2}z$, find a complex number which satisfies the equation $8w^4 + 3 = 2w(1 - w)$

9 It is given that $i\sqrt{5}$ is a root of the equation $P(z) = 0$, where $P(z) = z^4 + kz^2 + 5$ for k a real constant.

a Show that $k = 6$

b Find all the roots of the equation $P(z) = 0$

FP1

2.9 Complex roots and conjugate pairs

You can find the roots of the equation $z^2 - 2z + 5 = 0$ by using the quadratic formula.

$z^2 - 2z + 5 = 0$

so $\quad z = \frac{2 \pm \sqrt{-16}}{2} = \frac{2 \pm 4i}{2} = 1 \pm 2i$

The roots of this equation are $1 + 2i$ and $1 - 2i$.

You may notice that one root is the complex conjugate of the other. $(1 + 2i)^* = 1 - 2i$ and $(1 - 2i)^* = 1 + 2i$

If P(z) is a polynomial with real coefficients and $P(\alpha) = 0$ then $P(\alpha^*) = 0$

α^* is the complex conjugate of α. Refer to Section 2.4.

It follows from this that the roots of a polynomial equation with real coefficients occur in conjugate pairs.

EXAMPLE 1

Given that $2 + i$ is a root of the equation $P(z) = 0$ where $P(z) = z^3 - 2z^2 - 3z + 10$

a write down another root of $P(z)$

b completely factorise $P(z)$.

An exam question may use the term 'solution' in place of 'root'.

---

a All the coefficients of $P(z)$ are real numbers.
Hence $2 - i$ is also a root of the equation $P(z) = 0$

This follows since $(2 + i)^* = 2 - i$ and roots occur in conjugate pairs.

b $2 + i$ and $2 - i$ are roots of the equation $P(z) = 0$
Hence $(z - (2 + i))$ and $(z - (2 - i))$ are factors of $P(z)$.
$\therefore [z - (2 + i)][z - (2 - i)]$ is also a factor of $P(z)$

Factor theorem. Refer to C2

Expand the expression $[z - (2 + i)]\,[z - (2 - i)]$ and simplify:

$$\begin{aligned}
&[z - (2 + i)][z - (2 - i)] \\
&= z^2 - z(2 - i) - z(2 + i) + (2 + i)(2 - i) \\
&= z^2 - 2z + zi - 2z - zi + 4 - 2i + 2i - i^2 \\
&= z^2 - 4z + 5
\end{aligned}$$

The terms in zi cancel. $4 - i^2 = 4 + 1 = 5$

Hence $(z^2 - 4z + 5)$ is a factor of $P(z)$.

Divide P(z) by $z^2 - 4z + 5$ to find the remaining factor:

$$\begin{array}{r}
z + 2 \\
z^2 - 4z + 5 \overline{\big)\, z^3 - 2z^2 - 3z + 10} \\
\underline{z^3 - 4z^2 + 5z} \quad\;\; \\
2z^2 - 8z + 10 \\
\underline{2z^2 - 8z + 10} \\
0
\end{array}$$

Zero remainder confirms that $z^2 - 4z + 5$ is a factor of P(z).

The remaining factor of $P(z)$ is $(z + 2)$.

Hence $P(z) = (z^2 - 4z + 5)(z + 2)$
$= (z - (2 + i))(z - (2 - i))(z + 2)$

Make sure you give the complete factorisation of P(z).

You can find remaining factors by inspection rather than long division.

EXAMPLE 2

Given that $(x^2 + 2x + 3)$ is a factor of $P(x) = 2x^3 + 5x^2 + 8x + 3$, solve completely the equation $2x^3 + 5x^2 + 8x + 3 = 0$

To solve $P(x) = 0$, you need to find the remaining factor of $P(x)$.

Since $(x^2 + 2x + 3)$ is a quadratic factor of the cubic $P(x)$, the remaining factor must be a linear expression, say $(ax + b)$, for constants a and b to be determined.

linear × quadratic = cubic

$$2x^3 + 5x^2 + 8x + 3 \equiv (ax + b)(x^2 + 2x + 3)$$

$$\equiv ax^3 + \ldots + 3b$$

In this question, you only need to consider the first and last terms in the expansion in order to find *a* and *b*.

Equate coefficients of x^3:

$a = 2$

The identity symbol ≡ means that corresponding coefficients must be equal.

Equate constant terms:

$b = 1$

$3 = 3b$ so $b = 1$

Hence $2x^3 + 5x^2 + 8x + 3 \equiv (2x + 1)(x^2 + 2x + 3)$

If $2x^3 + 5x^2 + 8x + 3 = 0$
then $(2x + 1)(x^2 + 2x + 3) = 0$
so either $2x + 1 = 0$ or $x^2 + 2x + 3 = 0$

The equation $2x + 1 = 0$ has solution $x = -\frac{1}{2}$

The equation $x^2 + 2x + 3 = 0$ has roots given by

Discriminant $= 2^2 - 4 \times 1 \times 3$
$= -8$

$$x = \frac{-2 \pm \sqrt{-8}}{2} = \frac{-2 \pm i\sqrt{8}}{2}$$

$$= -1 \pm i\sqrt{2}$$

$\sqrt{-8} = 2i\sqrt{2}$

Hence the equation $2x^3 + 5x^2 + 8x + 3 = 0$ has roots

$x = -\frac{1}{2}$, $x = -1 \pm i\sqrt{2}$.

FP1

Exercise 2.9

1 Use an appropriate method to find the remaining factor of the given polynomial.

a $P(z) = z^3 + z^2 + 8z - 10$, given that $(z^2 + 2z + 10)$ is a factor of $P(z)$.

b $Q(z) = z^3 - 2z^2 - 13z - 10$, given that $(z^2 - 4z - 5)$ is a factor of $Q(z)$.

c $R(z) = 2z^3 - 9z^2 + 14z - 5$, given that $(z^2 - 4z + 5)$ is a factor of $R(z)$.

2 $1 - 2i$ is a root of the equation $P(z) = 0$, where
$P(z) = z^3 - z^2 + 3z + 5$

a Write down another complex root of this equation.

b Factorise $P(z)$ completely.

3 In each case α is a root of the equation $P(z) = 0$
Use this root to find all the factors of $P(z)$.

a $\alpha = 3 + i$, $P(z) = z^3 - 7z^2 + 16z - 10$

b $\alpha = 2 - i$, $P(z) = 2z^3 - 9z^2 + 14z - 5$

c $\alpha = 2 + 3i$, $P(z) = z^3 - 3z + 52$

4 Solve completely

a $2x^3 - 15x^2 + 26x + 17 = 0$ given that $4 - i$ is a root

b $x^3 - 8x + 32 = 0$ given that $2 - 2i$ is a root.

5 Given that $2i$ is a root of the equation $P(z) = 0$, where
$P(z) = z^4 - 3z^3 - 12z - 16$

a write down another root and hence show that $(z^2 + 4)$ is a factor of $P(z)$

b completely factorise $P(z)$ and hence solve the equation $P(z) = 0$

6 In each case, use the given root, α, to solve the equation.
Give your answers in the form $a + bi$ where $a, b \in \mathbb{R}$.

a $\alpha = 1 - 2i$, $x^4 - 2x^3 + x^2 + 8x - 20 = 0$

b $\alpha = 2 + i$, $x^4 - 4x^3 + 9x^2 - 16x + 20 = 0$

c $\alpha = 2 + 3i$ $x^4 - x^3 + 3x^2 + 31x + 26 = 0$

d $\alpha = -3 - i$ $x^4 - 16x^2 + 100 = 0$

7 It is given that $1 + 2i$ is a root of the equation $z^3 + z + k = 0$, where $k \in \mathbb{R}$.

a Simplify $(1 + 2i)^3$ and hence show that $k = 10$

b Solve completely the equation $z^3 + z + k = 0$

8 Given that $i\sqrt{2}$ is a root of the equation $P(z) = 0$, where
$P(z) = 2z^3 + 3z^2 + hz + k$, for real constants h and k,

a find the value of h and the value of k

b completely factorise $P(z)$.

FP1

9 Points A, B and C on the Argand diagram represent the three roots of the equation

$$z^3 - 7z^2 + 15z - 25 = 0$$

One of the roots of this equation is $1 + 2i$.

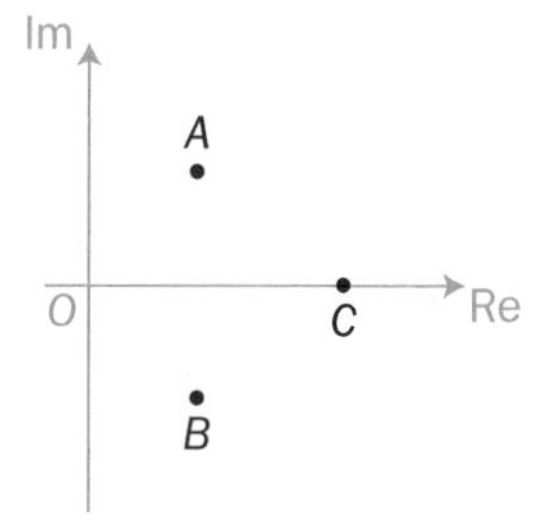

a Find each of the numbers represented by the points B and C.

b Prove that triangle OAC is right-angled.

c Describe fully the quadrilateral $OACB$ and find its area.

10 $2 - i$ is a root of the equation $z^3 - 11z + 20 = 0$

a Find the other two roots of the equation.

b On a single Argand diagram, illustrate the three roots of this equation.

c Show that the perimeter of the triangle with vertices defined by these three roots is $2(1 + \sqrt{37})$.

11 It is given that $1 + i$ is a root of the equation

$$x^4 - 2x^3 + 4x^2 - 4x + 4 = 0$$

a Write down another root of this equation.

b Find all the roots of this equation, giving each answer in the form $a + bi$ where $a, b \in \mathbb{R}$.

c Illustrate, on a single Argand diagram, the four points which represent the roots of this equation.

d Give a complete description of the quadrilateral with vertices defined by these points and calculate its exact area.

12 Prove that if the polynomial $P(z)$ has real coefficients and if α is a complex root of the equation $P(z) = 0$ then $(z^2 - 2z\text{Re}(\alpha) + |\alpha|^2)$ is a factor of $P(z)$.

13 a By using the fact that its roots occur in conjugate pairs, or otherwise, prove that any cubic equation with real coefficients must have at least one real root.

b Give an example of a cubic equation which has

i exactly one real root

ii no real roots.

FP1

1 Solve these equations.
Give answers in simplified surd form where appropriate.

a $2z^2 - 5z + 4 = 0$

b $\frac{5}{2 - z} = 3z$

c $\frac{z - 4}{2z + 1} = 2z$

d $\frac{1}{z - 1} + z = 1$

2 It is given that $z = 3 + 4i$ and $w = 2 - i$

a Express these quantities in the form $a + bi$, for $a, b \in \mathbb{R}$.

i $2z - 3w$

ii $\frac{1}{w - z}$

iii $(z + w)^2$

iv $\frac{1}{z} + \frac{i}{w}$

b Show that $w^2 = z^*$ and hence state the value of arg $(z - w^2)$.

3 By making the substitution $z = a + ib$, for a and b real numbers, solve the equation $2z + iz^* = 3 + 5i$

4 The non-zero complex number z is such that $\frac{z + i}{z - i} = \lambda$, where $\lambda \in \mathbb{R}$ is real.

a Show that z is purely imaginary and write down its imaginary part in terms of λ.

b Hence write down, in terms of π, the argument of z when
i $\lambda > 1$
ii $0 < \lambda < 1$

5 Solve the equation $\frac{2z + 1}{z} = 3 - i$

6 In modulus-argument form, the complex number $z = 12\left(\cos\frac{\pi}{6} + i\sin\frac{\pi}{6}\right)$

The complex number w is such that $\frac{z}{w} = 1 + i\sqrt{3}$

a Express z in exact cartesian form and hence show that $w = 3(\sqrt{3} - i)$

b Mark, on the same Argand diagram, points B and C representing the numbers w and $\frac{z}{w}$ respectively.

c Show that triangle OBC is right-angled and hence find the length BC, giving your answer in simplified surd form.

d Hence, or otherwise, show that $|w^2 - z| = 12\sqrt{10}$

7 On the Argand diagram, point P represents the complex number $z = 2 + 2i$
The line OQ is the result of rotating the line OP through $\frac{7}{12}\pi$ radians anticlockwise.

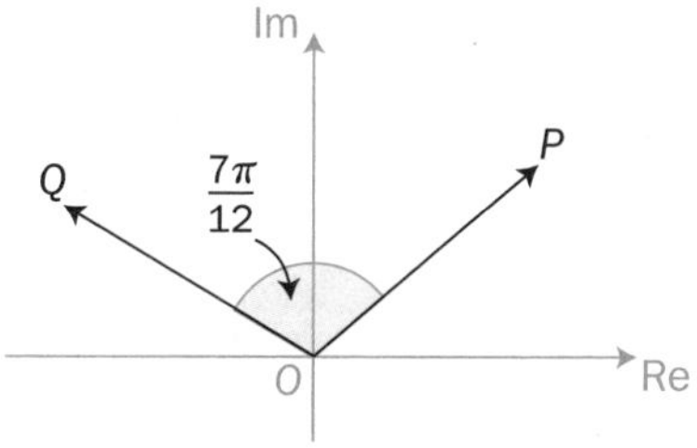

a Find the exact modulus and argument of z.

b Hence find the complex number represented by point Q.
Give your answer in exact cartesian form.

8 It is given that $z = \frac{ki}{\sqrt{3} + i}$, where $k < 0$ is a constant.

a Express z in exact cartesian form, giving real and imaginary parts in terms of k.

b Show that $|z| = -\frac{1}{2}k$ and find arg z, giving your answer in terms π.

c Show that z^3 is a real number and determine its sign.

9 Given that $2 + i$ is a root of the equation $P(z) = 0$, where

$P(z) = 2z^3 - 5z^2 - 2z + 15$

a write down another complex root of this equation

b solve the equation $P(z) = 0$.

FP1

10 It is given that $1 + 2i$ is a root of the equation $z^3 + az + 10 = 0$, where a is a real constant.

a Show that $a = 1$.

b Solve the equation $z^2 + \frac{10}{z} + 1 = 0$ for $z \neq 0$.

11 It is given that $2(1 + i\sqrt{3})$ is a root of the equation $z^3 + 64 = 0$

a Find the other two roots of this equation.

b Represent these three roots as points on a single Argand diagram.

c Show that these points lie on a common circle and state its radius.

d Give a geometrical reason why the sum of the roots of this equation is zero.

12 On the Argand diagram, points A, B and C represent the three roots of the equation

$$x^3 + 5x^2 + 11x + 15 = 0$$

Point D on the real axis is such that angle $CBD = \frac{1}{2}\pi$.

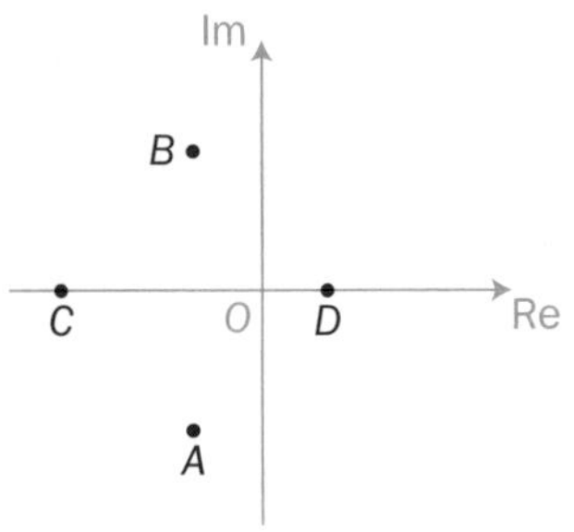

It is given that $-1 - 2i$ is a root of this equation.

a Find the other roots of the equation.
Hence write down each of the complex numbers represented by point A, point B and by point C.

b Show that triangle ABC is right-angled.

c Show that $OD = 1$ and hence find a polynomial equation of degree 4 whose roots on the Argand diagram are represented by the points A, B, C and D. Give your answer as a series of descending powers of x.

13 Given that $z = 22 + 4i$ and $\frac{z}{w} = 6 - 8i$, find

a w, in the form $a + bi$, where a and b are real

b the argument of z, in radians to two decimal places. [(c) Edexcel Limited 2002]

FP1

14 a Find the roots of the equation

$$z^2 + 2z + 17 = 0$$

giving your answers in the form $a + ib$, where a and b are integers.

b Show these roots on an Argand diagram. [(c) Edexcel Limited 2007]

15 The complex numbers z and w satisfy the simultaneous equations

$$2z + iw = -1$$
$$z - w = 3 + 3i$$

a Use algebra to find z, giving your answers in the form $a + ib$, where a and b are real.

b Calculate $\arg z$, giving your answer in radians to two decimal places. [(c) Edexcel Limited 2006]

16 The complex number $z = a + ib$, where a and b are real numbers, satisfies the equation

$$z^2 + 16 - 30i = 0$$

a Show that $ab = 15$

b Write down a second equation in a and b and hence find the roots of

$$z^2 + 16 - 30i = 0$$

[(c) Edexcel Limited 2004]

17 Given that $z = 2 - 2i$ and $w = -\sqrt{3} + i$

a find the modulus of wz^2.

It is given that $\arg(wz^2) = \frac{1}{3}\pi$

b Show on an Argand diagram the points A, B and C which represent z, w and wz^2 respectively, and determine the size of angle BOC.

18 Given that $3 - 2i$ is a solution of the equation

$$x^4 - 6x^3 + 19x^2 - 36x + 78 = 0,$$

a solve the equation completely.

b Show on a single Argand diagram the four points that represent the roots of the equation. [(c) Edexcel Limited 2006]

2 Exit

Summary

Refer to

- Complex numbers have the form $z = a + bi$, where a and b are real numbers and i is the imaginary number $\sqrt{-1}$
  $a = \text{Re}(z)$ is the real part of z, and $b = \text{Im}(z)$ is the imaginary part of z — 2.1, 2.2
- If $z = a + bi$ and $w = c + di$ then $z \pm w = (a + c) \pm (b + d)i$ — 2.3
- To divide z by w, multiply both top and bottom of $\frac{z}{w}$ by w^* — 2.3
- The conjugate of $z = a + bi$ is $z^* = a - bi$ — 2.4
- You can represent any complex number by a point, or by a line drawn from O, on an Argand diagram — 2.5
- The modulus of $z = a + bi$, written $|z| = \sqrt{a^2 + b^2}$, is the length r of the line representing z — 2.6
- An argument of $z = a + bi$, written $\arg z$, is an angle θ between a line representing z and the positive real axis. Angles measured clockwise are defined to be negative — 2.6
- An argument is principal if $-\pi < \arg z \leqslant \pi$ — 2.6
  Every non–zero complex number has a unique principal argument.
- Any complex number has modulus-argument form $z = r(\cos\theta + i\sin\theta)$ where $r = |z|$ and θ is an argument of z — 2.7
- $|zw| = |z||w|$ $\quad \left|\frac{z}{w}\right| = \frac{|z|}{|w|}$ $\quad$ (for $w \neq 0$) — 2.7
- if α is a root of the polynomial equation $\text{P}(z) = 0$, then $\text{P}(\alpha) = 0$ — 2.8
- If $\text{P}(z)$ is a polynomial with real coefficients and $\text{P}(\alpha) = 0$ then $\text{P}(\alpha^*) = 0$ — 2.9

FP1

Links

Complex numbers are very important in analysing many aspects of the physical world.
Equations involving complex numbers can be used to predict the motion of electronic particles allowing engineers to design integrated circuits. These circuits are central to the development of much modern technology such as cars, computers and mobile phones.

Many of today's blockbuster films would not be possible without modern mathematics. Animation techniques rely on complex numbers called quarternions (these are numbers of the form $q = a_0 + a_1i + a_2j + a_3k$ where $i^2 = j^2 = k^2 = ijk = -1$ and a_0, a_1, a_2 and a_3 are real numbers) which can be used to represent rotations in 3-dimensional space.

3 Numerical solution of equations

This chapter will show you how to

- use the methods of interval bisection and linear interpolation to locate a root of an equation
- use the Newton-Raphson procedure to produce a sequence of iterates which converge to a root of an equation.

Before you start

You should know how to:

1 Find the equation of the straight line passing through two given points.

e.g. Find the equation of the straight line passing through the points $A(3, 6)$ and $B(7, -2)$.

Gradient $m = \frac{y_2 - y_1}{x_2 - x_1} = \frac{(-2) - 6}{7 - 3}$

$= -2$

The line has equation $y = mx + c$

Substitute for m, x and y (using point A):

$6 = -2 \times 3 + c$

$c = 12$

Hence the equation of the line is $y = -2x + 12$

2 Differentiate standard functions

e.g. Find $f'(x)$ where $f(x) = 2\sqrt{x}(x + 1)$

$f(x) = 2x^{\frac{1}{2}}(x + 1)$

$= 2x^{\frac{3}{2}} + 2x^{\frac{1}{2}}$

$f'(x) = 3x^{\frac{1}{2}} + x^{-\frac{1}{2}}$

3 Calculate values of an iterative sequence.

e.g. Find the value of x_1 and x_2 for the sequence defined by the iterative formula

$$x_{n+1} = 1 + \frac{3}{x_n},\ x_0 = 1$$

$x_0 = 1$, $x_1 = 1 + \frac{3}{1} = 4$, $x_2 = 1 + \frac{3}{4} = \frac{7}{4}$

Hence $x_1 = 4$, $x_2 = 1.75$

Check in:

See C1 for revision.

1 a Find the equation of the line passing through the points $A(4, 24)$ and $B(-3, 3)$.

b The line L passes through the points $A(2, 4)$ and $B(8, 1)$. Find the co-ordinates of the point P where L crosses the x-axis.

2 Find $f'(x)$ where Refer to 1.5.

a $f(x) = x(6\sqrt{x} - 5)$

b $f(x) = \frac{x^2 - 1}{x}$

c $f(x) = \frac{1}{x} - \frac{4}{\sqrt{x}}$

3 Find the value of x_1 and x_2 for the sequence defined by the iterative formula Refer to 1.4.

a $x_{n+1} = \frac{x_n^2}{3 + x_n}$, $x_0 = 3$

b $x_{n+1} = \frac{4x_n}{\sqrt{1 + x_n}}$, $x_0 = 15$

3.1 Intervals and roots

An interval is a section of the real number line.

You can describe a range of numbers on the x-axis using interval notation.

An interval contains all real values between its end points.
A closed interval $[a, b]$ includes the end points a and b.
An open interval (a, b) does not include the end points.

e.g. This diagram shows the interval [2, 5]. 2 and 5 are the end points of the interval.

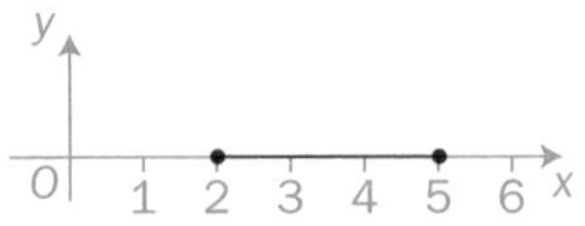

EXAMPLE 1

a Give an example of a real number in the interval [3.5, 4.5].

b State the width of the interval (1, 4.2) and list all the integers that it contains.

---

a The interval [3.5, 4.5] contains all real numbers x such that $3.5 \leqslant x \leqslant 4.5$.
$x = 3.7$ is an example of a number in this interval.

[3.5, 4.5] includes the end points 3.5 and 4.5.

b The interval (1, 4.2) contains all real numbers x such that $1 < x < 4.2$.
The width of this interval is $4.2 - 1 = 3.2$
The integers in this interval are 2, 3, and 4.

(1, 4.2) does not include the end points 1 and 4.2.

To find the width of an interval, subtract the lower end point from the upper end point.

A root of an equation $f(x) = 0$ is any value α for which $f(\alpha) = 0$

See C2 for revision.

Sometimes you can find the exact roots of an equation of the form $ax^2 + bx + c = 0$ by using the quadratic formula,

$$x = \frac{-b \pm \sqrt{b^2 - 4ac}}{2a}$$

e.g. The roots of the equation $x^2 + 4x - 2 = 0$ are given by

$$x = \frac{-4 \pm \sqrt{24}}{2} = -2 \pm \sqrt{6}$$

When an exact solution of an equation does not exist, or is very difficult to find, you can use a numerical method to approximate its roots.

You can estimate a root of an equation by showing that it lies in an interval (a, b) whose end points a and b are known.

If a function $f(x)$ is continuous over an interval (a, b), and the values $f(a)$ and $f(b)$ have opposite signs, then the equation $f(x) = 0$ has a root in the interval (a, b).

Continuous means there are no breaks in the graph of the function.

You can determine the value of a given root of an equation $f(x) = 0$ to any required level of accuracy by showing that $f(x)$ changes sign over an appropriate interval.

EXAMPLE 2

a Show that the equation $x^3 - 2x - 3 = 0$ has a root α in the interval (1, 2).

b Given that α is the only root of this equation, show that $\alpha = 1.89$ to two decimal places.

---

a $f(x) = x^3 - 2x - 3$ is a continuous function.

Evaluate the function at each end point of the interval (1, 2):

$f(1) = 1^3 - 2 \times 1 - 3$
$= -4 < 0$

$f(2) = 2^3 - 2 \times 2 - 3$
$= 1 > 0$

Since $f(x) = x^3 - 2x - 3$ changes sign in the interval (1, 2), the equation $x^3 - 2x - 3 = 0$ has a root between $x = 1$ and $x = 2$, as required.

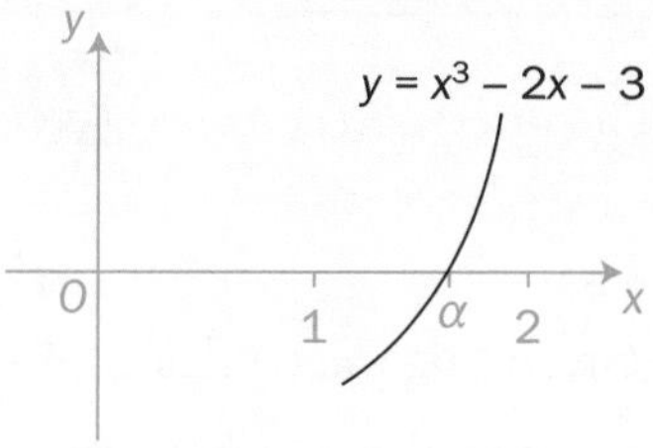

This function 'changes sign' in the interval (1, 2).

b Show $\alpha = 1.89$ to two decimal places by proving that α lies in the interval (1.885, 1.895).
A change of sign in this interval will prove that it contains α.

Any value in the interval (1.885, 1.895), when rounded to 2 d.p. equals 1.89.

Test the sign of f(x) at each end point of this interval:

$f(1.885) = 1.885^3 - 2 \times 1.885 - 3$
$= -0.0721\ldots < 0$

$f(1.895) = 1.895^3 - 2 \times 1.895 - 3$
$= 0.0149\ldots > 0$

Since $f(x)$ changes sign in (1.885, 1.895), the unique root α of the equation $f(x) = 0$ must lie in this interval and hence $\alpha = 1.89$ to two decimal places.

Exercise 3.1

1 Use the change of sign method applied to a suitable function to show that each equation has a root α in the specified interval.

a $3^x - 3x - 4 = 0, \quad (2, 3)$

b $x + \log_{10}(x) - 2 = 0, \quad [1, 2]$

c $\dfrac{3}{\cos x} - x^2 = 0, \quad (4.75^c, 4.85^c)$

d $x^2 - 2^{0.5x} = 0, \quad [-1, 0]$

2 **a** Show that the equation $2^x - x - 3 = 0$ has a root α in the interval (2, 3).

b Given that α is the only root in this interval, prove, by means of a change of sign in an appropriate interval, that $\alpha = 2.44$ to two decimal places.

3 A particular equation has a root α where $\alpha = 0.708$ to three decimal places.
Write down an interval of minimum width which must contain α.
State the width of this interval.

4 Consider the equation $f(x) = 0$, where $f(x) = 4x^2 - 4x + 1$

a Verify that

i $f(x)$ does *not* change sign at the end points of the interval (0, 1)

ii the equation $f(x) = 0$ has a root $\frac{1}{2}$.

b Explain why part **a** does not contradict the 'change of sign' principle.

FP1

3.2 Interval bisection

You can find a more accurate estimate for a root α of an equation by bisecting an interval which is known to contain α.

Bisect means to cut in half.

The intervals (a, b) and $[a, b]$ each have midpoint $\frac{a+b}{2}$.

EXAMPLE 1

Given that the equation $2^x - 5x - 3 = 0$ has exactly one positive root, α,

a show that α is between $x = 4.6$ and $x = 4.8$

b starting with the interval (4.6, 4.8), use interval bisection to find an estimate for α which is correct to one decimal place.

---

a Define $f(x) = 2^x - 5x - 3$

Defining a function such that $f(\alpha) = 0$ makes it easier to apply the method.

Find the sign of the function at each end point of the interval (4.6, 4.8):

$f(4.6) = -1.748\ldots < 0$ and $f(4.8) = 0.857\ldots > 0$

$f(x)$ changes sign in the interval (4.6, 4.8)
$\therefore f(x) = 0$ has a root between $x = 4.6$ and $x = 4.8$

Refer to Section 3.1.

Hence α is between $x = 4.6$ and $x = 4.8$

b The midpoint of the interval (4.6, 4.8) is $x = 4.7$

$x = 4.7$ bisects the interval [4.6, 4.8].

The sign of $f(x)$ at $x = 4.7$ determines in which of the intervals (4.6, 4.7) and (4.7, 4.8) α lies.

Find f(4.7):

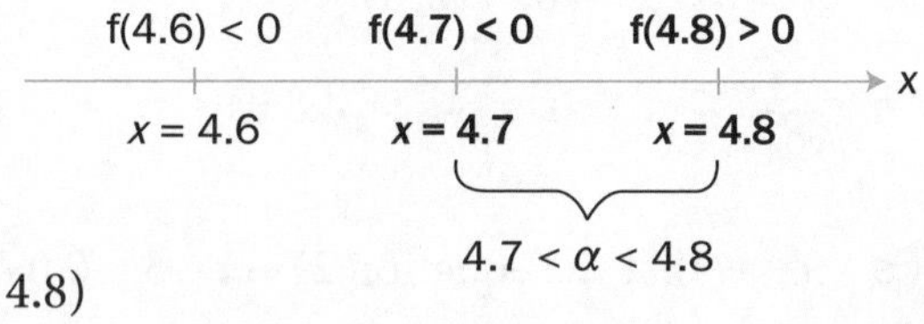

$f(4.7) = 2^{4.7} - 5 \times 4.7 - 3$
$= -0.507\ldots < 0$

The change of sign occurs in the interval (4.7, 4.8)
Hence α is between $x = 4.7$ and $x = 4.8$

To find α to 1 d.p. apply interval bisection to the interval (4.7, 4.8):

The midpoint of the interval (4.7, 4.8) is $x = 4.75$

Midpoint = $(4.7 + 4.8) \div 2$

The sign of $f(x)$ at $x = 4.75$ determines α to 1 d.p.

Find f(4.75):

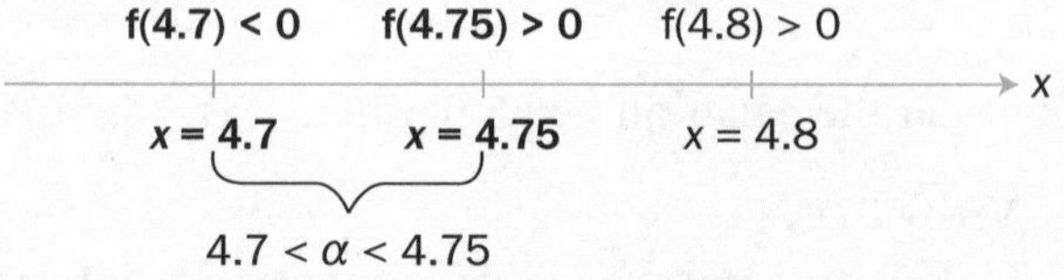

$f(4.75) = 2^{4.75} - 5 \times 4.75 - 3$
$= 0.158\ldots > 0$

Hence, to one decimal place, $\alpha = 4.7$

Any number in the interval (4.7, 4.75) has value 4.7 when rounded to one decimal place.

Exercise 3.2

Where appropriate, you may assume the root in the given interval is unique.

1 Given that $f(x) = 0$ has a root α in the specified interval, apply interval bisection to the given interval to find α correct to one decimal place.

a $f(x) = x^3 - x - 1, \quad (1.2, 1.4)$

b $f(x) = x^2 - 2^{-x}, \quad (0.6, 0.8)$

c $f(x) = 4 - x\log_{10} x, \quad (5.4, 5.5)$

d $f(x) = 3^{-x}x + 5, \quad (-1.4, -1.2)$

2 $f(x) = 2^x - \frac{1}{x}$, where $x > 0$

a Show that the equation $f(x) = 0$ has a root α in the interval $(0.64, 0.66)$.

b Hence, starting with the interval $(0.64, 0.66)$, use interval bisection to find α correct to two decimal places.

3 In each case, the equation $f(x) = 0$, where x is measured in radians, has a root α in the specified interval.
Apply interval bisection to the given interval to find α correct to one decimal place.

a $f(x) = x\sin x - 1 \qquad [2.6, 2.8]$

b $f(x) = x^2 - \cos x \qquad [0.7, 0.9]$

4 a Show that the equation

$$x^2 - 30 = 0$$

has a root α in the interval $[5.4, 5.5]$.

b Starting with the interval $[5.4, 5.5]$, use interval bisection three times to find an interval of width 0.0125 which contains $\sqrt{30}$.

5 a On the same diagram, sketch the graphs with equations

$$y = 2^x \quad \text{and} \quad y = \frac{5}{x}$$

b Deduce that the equation

$$2^x x - 5 = 0$$

has precisely one root, α.

c Given that α is in the interval $(1.5, 1.8)$, use interval bisection starting with this interval to find α correct to one decimal place.

6 $f(x) = \log_{10} x + 5x - 7, x > 0$

a On the same diagram, sketch the graph with equation

i $y = \log_{10} x$

ii $y = 7 - 5x$

Label the point where each graph crosses the x-axis with its coordinates.

b Deduce that the equation $f(x) = 0$ has exactly one real root α, where $1 < \alpha < 1.4$

c Starting with this interval, use interval bisection to find α to 1 decimal place.

7 Let $f(x) = x^x - 6$ for $x > 0$

a Show that $f(x) = 0$ has a root α in the interval (2.2, 2.4).

b Starting with the interval (2.2, 2.4) use interval bisection to find an interval of width 0.025 which contains α and hence state the value of α to as many decimal places as can be justified.

c Show that one further application of interval bisection does not improve the accuracy of the estimate for α found in part **b**.

8 $f(x) = x^3 + 1 - \frac{1}{x}, x \neq 0$

a Show, by means of a sketch, that the equation $f(x) = 0$ has two real roots $\alpha < \beta$

It is given that α lies in the interval [–2, –1]

b Use interval bisection starting with this interval to find an interval of width 0.25 which contains α.

c Show, by means of a change of sign, that β lies in the interval [0.5, 1].

d Apply the method of interval bisection starting with the interval [0.5, 1] to find β to 1 decimal place.

e Show that the method of interval bisection fails to locate a root of the equation $f(x) = 0$ when applied to the interval [–1, 1]. Explain the cause of this failure.

9 The diagram shows the graph of the equation $y = f(x)$. You may assume that the values α, β and γ where the graph crosses the x-axis are accurately positioned.

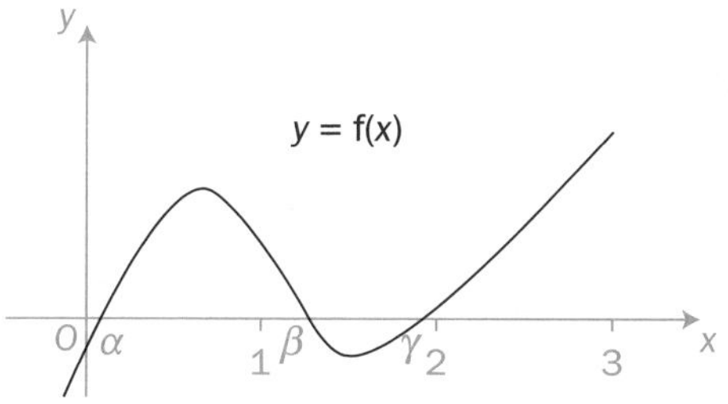

a State to which of the three roots the method of interval bisection will converge when applied to the starting interval (0, 2).

b Explain why (1, 3) is an unsuitable starting interval when using interval bisection to estimate the roots of the equation $f(x) = 0$.

10 It is given that $7.93 < \sqrt[3]{500} < 7.95$.
Use interval bisection over a suitable interval to obtain an estimate for $\sqrt[3]{500}$ which is correct to two decimal places.

11 Let $f(x) = x^2 - \tan\left(\frac{1}{2}x\right) + 1$, for $0 \leqslant x \leqslant 2\pi$, $x \neq \pi$.

a Show that the equation $f(x) = 0$ has a root, α, between $x = 2.9$ and $x = 3.1$

b Starting with [2.9, 3.1], use interval bisection to find α to one decimal place.

c Show that the method of interval bisection fails when applied to the interval [2.9, 3.4].
Explain the cause of this failure.

12 Let $f(x) = \cos(2x) - 2^{-x}$, where x is in radians.

a On the same diagram, sketch the graphs with equations
$y = \cos 2x$ and $y = 2^{-x}$

b Deduce that the equation $f(x) = 0$ has
i an infinite number of positive roots
ii no negative roots.

The smallest positive root α of the equation $f(x) = 0$ lies between $x = 0.32$ and $x = 0.34$

c Starting with the interval (0.32, 0.34) use interval bisection to find α correct to two decimal places.

d With reference to your sketch only, show that $\alpha + \beta < \pi$, where β is the next smallest positive root of the equation $f(x) = 0$

3.3 Linear interpolation

You can estimate a root of an equation $f(x) = 0$ by assuming that the graph of $y = f(x)$ is a straight line over an interval which contains the root.

EXAMPLE 1

The diagram shows part of the graph with equation $y = f(x)$, where $f(x) = x^2 - 4^{-x}$

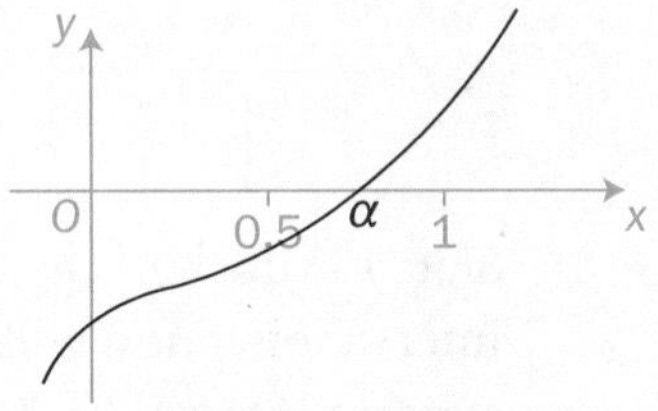

The equation $f(x) = 0$ has a root, α, where $0.5 \leqslant \alpha \leqslant 1$.
Use linear interpolation on the values at the end points of the interval $[0.5, 1]$ to find an approximation to α.

Evaluate the function at each of the end points of the interval [0.5, 1]:

$f(0.5) = -0.25 \quad f(1) = 0.75$

The change of sign is a good check that your values are correct.

Approximate the curve by a straight line joining the points $A(0.5, -0.25)$ and $B(1, 0.75)$:

The value x_1 where the line AB crosses the x-axis is an estimate for α.

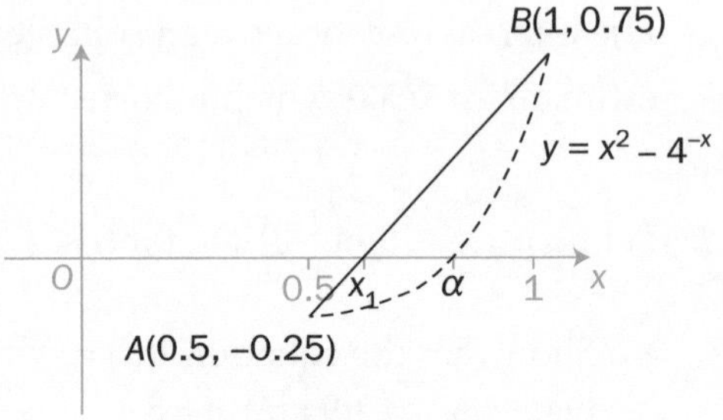

Find x_1 by first finding the equation of the line AB:

The line AB has gradient $\text{Grad}_{AB} = \dfrac{0.75 - (-0.25)}{1 - 0.5} = 2$

Hence AB has equation $y = 2x + c$

Use the point $B(1, 0.75)$ to give the equation of line AB:

$y = 2x + c$, so $0.75 = 2 \times 1 + c$
$c = -1.25$

$$y = 2x - 1.25$$

The line AB crosses the x-axis when $2x - 1.25 = 0$

so $x = 0.625$

$2x - 1.25 = 0$,
$x = \frac{1}{2} \times 1.25 = 0.625$

Hence, by linear interpolation, $x_1 = 0.625$ is an approximation to α.

Exercise 3.3

1 Each equation $f(x) = 0$ has a root, α, in the specified interval.
Using the end points of each interval find, by linear interpolation, an estimate for α.

a $f(x) = x^3 - x - 3 \quad [1, 2]$

b $f(x) = 2^{2x-1} - 3x \quad [1.5, 2]$

c $f(x) = 8x^2 - 1 + 3\sqrt{x} \quad (0, 0.25)$

d $f(x) = \sin(\pi x) - 2x + 3 \quad (1^c, 1.5^c)$

2 Given that $f(x) = x^3 - 3x + 1$

a show that $f(x) = 0$ has a root, α, between $x = 1$ and $x = 2$

b using the end points of the interval $(1, 2)$ find, by linear interpolation, an estimate for α.

3 Given that $f(x) = x^2 - \sqrt{x} - 4$, where $x \geqslant 0$

a show, by means of a sketch, that $f(x) = 0$ has exactly one root, α

b show that α lies in the interval (2.3, 2.4)

c using the end points of the interval, find, by linear interpolation, an estimate for α. Give your estimate correct to two decimal places.

4 Let $f(x) = x\,3^{x-1} - 4$

It is given that the equation $f(x) = 0$ has exactly one root, α.

a Show that α lies in the interval [1.7, 1.8].

b Find an estimate, x_1, for α by applying the method of linear interpolation to the interval [1.7, 1.8]. Give your answer to two decimal places.

c Determine whether your answer to part **b** is an under- or over-estimate for α.

5 The diagram shows an accurate sketch of the graph of a function $f(x)$ for $c \leqslant x \leqslant d$. The graph crosses the x-axis at the point $x = \alpha$.

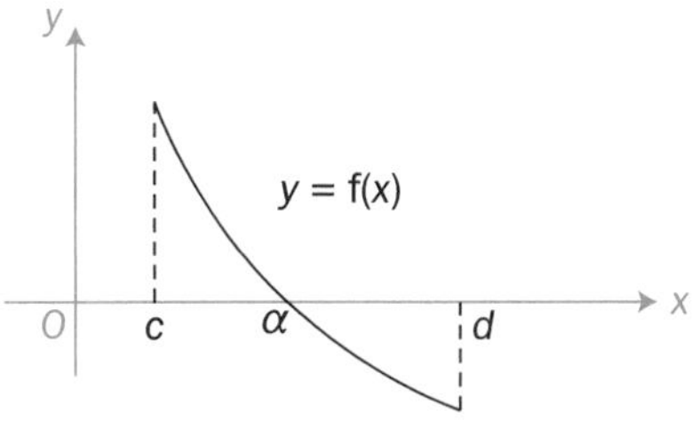

a Explain why the result of applying linear interpolation once to the interval $[c, d]$ produces an over-estimate for the root, α, of the equation $f(x) = 0$

b On separate diagrams, sketch an example of a situation where linear interpolation produces
i an under-estimate
ii an exact answer for a root, α, of an equation $f(x) = 0$

6 a Use a sketch to show that the equation $f(x) = 0$, where $f(x) = x^3 - 3^{-x} - 1$, has exactly one real root, α.

b Show, by calculation, that $1 < \alpha < 1.5$.

c By applying linear interpolation to the equation $f(x) = 0$ across the interval (1, 1.5), find an estimate for α. Give your answer to one decimal place.

d Show that α lies in the interval $(1, x_1)$, where x_1 is the rounded answer found in part **c**, and hence find a second estimate, x_2, for α using linear interpolation on the equation $f(x) = 0$ across this interval. Give your answer to two decimal places.

3.4 The Newton-Raphson process

You can use the derivative of the function $f(x)$ to obtain a very accurate estimate to a root of the equation $f(x) = 0$

You write the derivative of the function $f(x)$ as $f'(x)$. See C1 for revision.

The diagram shows the graph of the equation $y = f(x)$
It is given that $x = \alpha$ is a root of the equation $f(x) = 0$
and that $x = x_1$ is an estimate for α.

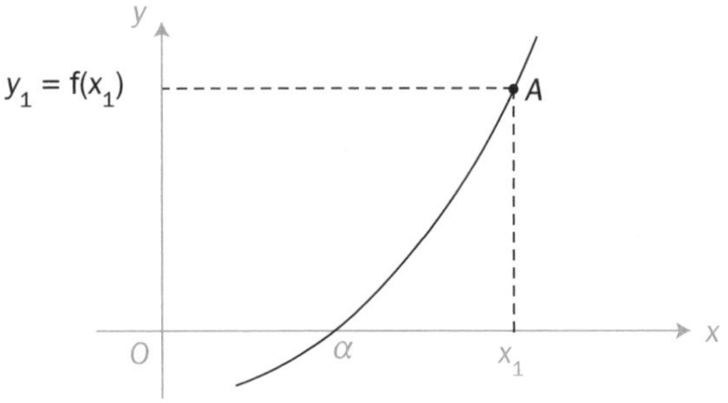

The tangent T to the curve at the point $A(x_1, y_1)$ crosses the x-axis at the point $B(x_2, 0)$.

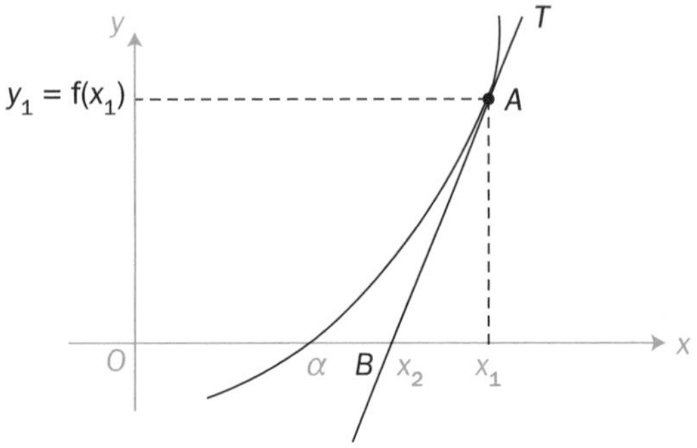

In general, x_2 will be a better estimate for α than x_1.

Since T passes through the points $A(x_1, f(x_1))$ and $B(x_2, 0)$,

its gradient is $\dfrac{f(x_1)}{x_1 - x_2}$

$y_1 = f(x_1)$

Hence $\dfrac{f(x_1)}{x_1 - x_2} = f'(x_1)$

$$\frac{f(x_1)}{f'(x_1)} = x_1 - x_2$$

$$x_2 = x_1 - \frac{f(x_1)}{f'(x_1)}$$

$f'(x_1)$ is the gradient of the tangent to the curve $y = f(x)$ at the point $x = x_1$

Applying this process to x_2 produces another estimate for α, x_3,

where $\quad x_3 = x_2 - \dfrac{f(x_2)}{f'(x_2)}$

In general, x_3 will be a better estimate for α than x_2.

FP1

In general, if the equation $f(x) = 0$ has a root $x = \alpha$, and $x = x_1$ is a first approximation to α, then the iterates $x_2, x_3, x_4, \ldots$ produced by the formula $x_{n+1} = x_n - \frac{f(x_n)}{f'(x_n)}$ are increasingly accurate approximations to α.

This result is in the formula booklet (under FP1).

Refer to C1 for revision on iterative sequences.

This is the Newton-Raphson method.
You can apply it to find the estimate of a root of a given equation.

EXAMPLE 1

It is given that the equation $x^3 = x^2 + 5$ is satisfied by a value, α, where $2 \leqslant \alpha \leqslant 3$.
Taking $x_1 = 2$ as a first approximation to α, apply the Newton-Raphson process to a suitable function twice to find second and third approximations to α.
Give each answer to three significant figures.

α is a root of the equation $f(x) = 0$,
where $f(x) = x^3 - x^2 - 5$

Rearrange the given equation into the form $f(x) = 0$ before using the method.

$$f(x) = x^3 - x^2 - 5$$
$$f'(x) = 3x^2 - 2x$$

The first approximation is given as $x_1 = 2$

Use $x_2 = x_1 - \frac{f(x_1)}{f'(x_1)}$ to calculate the second approximation:

$$f(x_1) = f(2) = 2^3 - 2^2 - 5$$
$$= -1$$

$$f'(x_1) = f'(2) = 3 \times 2^2 - 2 \times 2$$
$$= 8$$

Hence $x_2 = x_1 - \frac{f(x_1)}{f'(x_1)}$

$$= 2 - \frac{(-1)}{8}$$
$$= 2.125$$

so $x_2 = 2.13$ (3 s.f.)

Use $x_3 = x_2 - \frac{f(x_2)}{f'(x_2)}$ to calculate the third approximation:

Use the unrounded value of x_2 to calculate x_3.

$f(x_2) = f(2.125) = 0.0800\ldots$ and $f'(x_2) = f'(2.125) = 9.2968\ldots$

Hence $x_3 = 2.125 - \frac{0.0800\ldots}{9.2968\ldots}$

Do not round off the intermediate values.

$$= 2.11638\ldots$$

so $x_3 = 2.12$ to three significant figures

FP1

EXAMPLE 2

The diagram shows an accurate sketch of the graph with equation $y = \mathrm{f}(x)$, where $\mathrm{f}(x) = 8\sqrt{x} - x^3 - 2$ for $x \geqslant 0$. The graph crosses the x-axis exactly once, at $x = \alpha$.

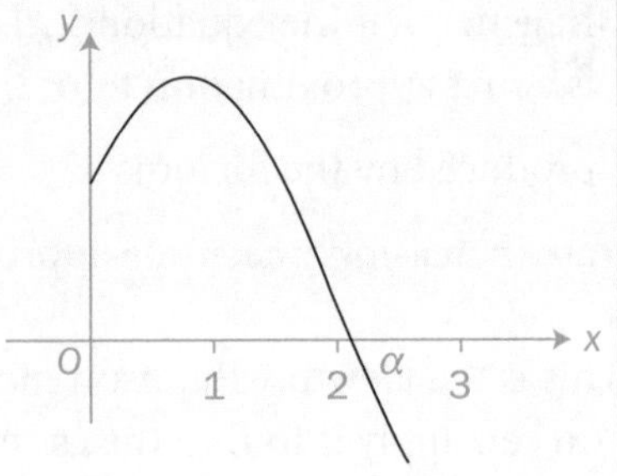

a Taking $x_1 = 2$ as a first approximation to α, apply the Newton-Raphson procedure once to $\mathrm{f}(x)$ to find a second approximation to α.

b By demonstrating a change of sign of $\mathrm{f}(x)$ across a suitable interval, show that, as an approximation to α, your answer to part **a** is accurate to one decimal place.

You can use a change of sign to show the accuracy achieved by the Newton-Raphson process.

c Use the sketch to explain why $x_1 = 1$ is an unsuitable initial value when applying the Newton-Raphson procedure to $\mathrm{f}(x)$ to approximate α.

---

a $\mathrm{f}(x) = 8x^{\frac{1}{2}} - x^3 - 2$, so $\mathrm{f}'(x) = 4x^{-\frac{1}{2}} - 3x^2$

$\sqrt{x} = x^{\frac{1}{2}}$

The first approximation is $x_1 = 2$

Use $x_2 = x_1 - \dfrac{\mathrm{f}(x_1)}{\mathrm{f}'(x_1)}$ to calculate the second approximation:

$$\mathrm{f}(x_1) = \mathrm{f}(2) = 8 \times 2^{\frac{1}{2}} - 2^3 - 2 = 1.3137\ldots$$

$$\mathrm{f}'(x_1) = \mathrm{f}'(2) = 4 \times 2^{-\frac{1}{2}} - 3 \times 2^2 = -9.1715\ldots$$

Hence $x_2 = x_1 - \dfrac{\mathrm{f}(x_1)}{\mathrm{f}'(x_1)} = 2 - \dfrac{1.3137\ldots}{(-9.1715\ldots)}$

$= 2.1432\ldots$

b To one decimal place, $x_2 = 2.1$

To show that the unique root α also equals 2.1 to 1 d.p, you must show that f(x) changes sign across the interval (2.05, 2.15).

Find f(2.05) and f(2.15):

$$\mathrm{f}(2.05) = 8 \times 2.05^2 - 2.05^3 - 2 = 0.839\ldots$$
$$\mathrm{f}(2.15) = 8 \times 2.15^2 - 2.15^3 - 2 = -0.208\ldots$$

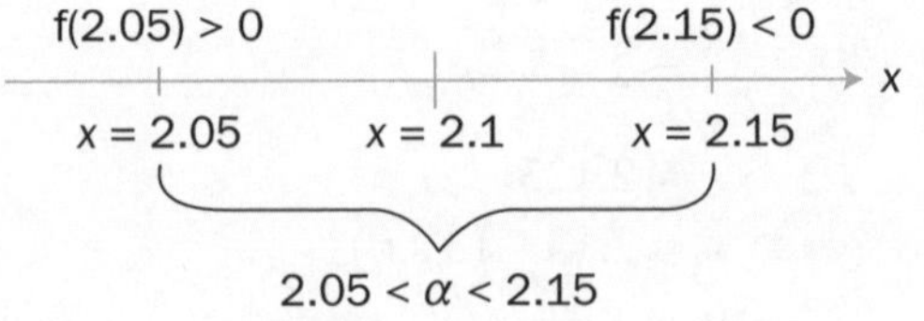

Any number in the interval (2.05, 2.15) has value 2.1 when rounded to 1 d.p.

The change of sign shows that $\alpha = 2.1$ to 1 d.p.

Hence, as an approximation to α, x_2 is accurate to 1 d.p.

c $x_1 = 1$ is near to the x-coordinate of the stationary point on the graph. Starting the procedure with $x_1 = 1$ produces a negative value for x_2. Since the curve is not defined for $x < 0$, the next tangent in the process cannot be drawn and the procedure breaks down.

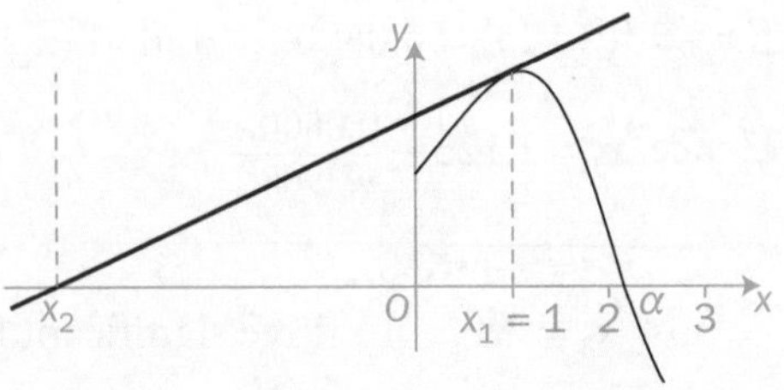

The Newton-Raphson process is based on drawing tangents to the curve $y = \mathrm{f}(x)$.

FP1

Starting near a stationary point can cause the Newton-Raphson process to become unstable.

Exercise 3.4

1 Each equation $f(x) = 0$ has a root α. Taking x_1 as a first approximation to α, use the Newton-Raphson process once on $f(x)$ to obtain a second approximation to α. Give your answers to two decimal places where appropriate.

a $f(x) = x^3 - 4x + 1$ $\quad x_1 = 2$

b $f(x) = x^3 - 2x^2 - x + 3$ $\quad x_1 = 1$

c $f(x) = x^4 - 4x^2 - 2x - 1$ $\quad x_1 = -2$

d $f(x) = x^2 - \dfrac{1}{x} - 3$ $\quad x_1 = -2$

e $f(x) = x^3 - 4\sqrt{x} + 2$ $\quad x_1 = 3$

2 In each of the following, the given equation is satisfied by a value α. Starting with x_1 as a first approximation to α, apply the Newton-Raphson process once to a suitable function to find a second approximation for α. Give your answers to three decimal places.

a $x^3 + 2x^2 - 11x = 13$ $\quad x_1 = 3$

b $x^2 = \dfrac{1}{x} + 2$ $\quad x_1 = 2$

c $x(x - 2) = 4 + \sqrt{x}$ $\quad x_1 = 4$

3 The equation $\dfrac{1}{\sqrt{x}} + 3x^2 - 3 = 0$ has exactly one root, α, where $0 < \alpha < 1$.

a Starting with $x_1 = 1$ as a first approximation to α, apply the Newton-Raphson process once to the function $f(x) = \dfrac{1}{\sqrt{x}} + 3x^2 - 3$ to find a second approximation for α giving your answer to 2 d.p.

b Show that, as an approximation to α, the value of x_2 found in part **a** is
i accurate to one decimal place
ii not accurate to two decimal places.

4 It is given that the equation $x^4 - 3x^2 - 7 = 0$ has a positive root, α, between $x = 2$ and $x = 3$

a Taking $x_1 = 2$ as a first approximation to α, use the Newton-Raphson process once on $f(x)$ to obtain a second approximation to α.

b Find, by solving an appropriate quadratic equation, the exact value of α.

c Hence determine the percentage error of the approximation to α found in part **a**.

5 The diagram shows the curve with equation $y = f(x)$ where $f(x) = 2x^3 - 11x^2 + 10x + 9$. The graph has three roots, α, β and γ, where $\beta < 3 < \alpha$

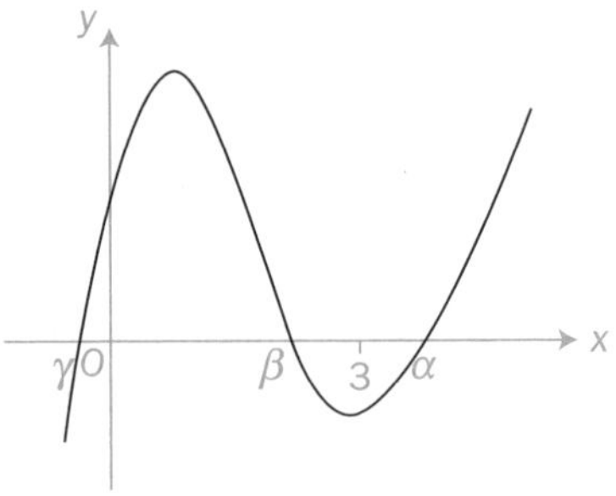

a Taking $x_1 = 4$ as a first approximation to α, use the Newton-Raphson process twice on $f(x)$ to obtain a second and third approximation to α.
Give your answers to three decimal places.

b Show that, as an approximation to α, the rounded value of x_3 found in part **a** is accurate to three decimal places.

c i Apply the Newton-Raphson process once to $f(x)$ with a starting value $x_1 = 3$ to find an approximation to one of the roots.

ii State the root to which further iterations of the process would converge.

6 $f(x) = \dfrac{x-1}{x^2} + x$ where $x > 0$

a Show that $f(x) = \dfrac{1}{x} - \dfrac{1}{x^2} + x$

and hence find an expression for $f'(x)$.

It is given that the equation $f(x) = 0$ has exactly one root α.

b Taking $x_1 = 0.5$ as a first approximation to α, use the Newton-Raphson process once on $f(x)$ to obtain a second approximation to α. Give your answer to two decimal places.

c Show by direct calculation that applying the Newton-Raphson process to $f(x)$ with $x_1 = 1$ as a first approximation to α is *less* efficient than starting it at $x_1 = 0.5$ to approximate this root.

7 The diagram shows the curve with equation $y = f(x)$ where $f(x) = \dfrac{x^2 + 4}{\sqrt{x}} - 6$, for $x > 0$. The graph has two roots, α and β.

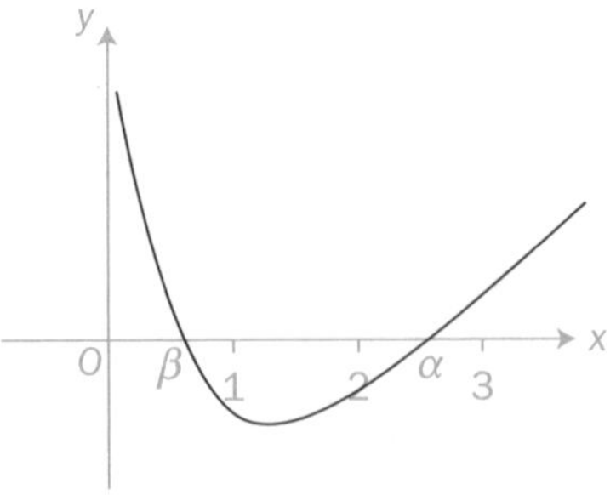

a Taking $x_1 = 3$ as a first approximation to α, use the Newton-Raphson process once on $f(x)$ to obtain a second approximation to α.
Give your answer to one decimal place.

b Show by direct calculation that applying the Newton-Raphson process to $f(x)$ with $x_1 = 1$ as a first approximation to β leads to an error.
Explain fully the cause of this error.

FP1

8 The function $f(x) = 2x^3 + kx - 1$, where k is a constant, is such that the equation $f(x) = 0$ has exactly one root α in the interval $(1, 2)$. A single application of the Newton-Raphson process to $f(x)$ with a starting value of $x_1 = 2$ produces a second approximation $x_2 = \frac{11}{6}$ to α.

a Show that $k = -6$.

b Continue the Newton-Raphson process to $f(x)$ to obtain a third approximation x_3 to α.

c Show that as an approximation to α the value obtained for x_3 in **b** is accurate to three decimal places.

d Explain why it would not be appropriate to apply the Newton-Raphson process to $f(x)$ with a starting value of $x_1 = 1$ in order to find an approximation to α.

9 In this question you may assume that the derivative of $\sin\theta$ with respect to θ is $\cos\theta$.

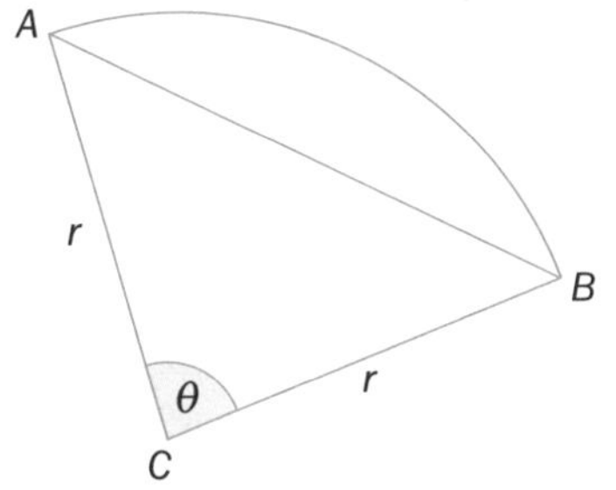

The diagram shows a sector of a circle of radius r. Points A and B lie on the circle and subtend an angle θ at the centre, C.

a Given that the area of triangle ABC is half the area of sector ABC, show that $2\sin\theta - \theta = 0$

b Taking $\theta_1 = 2$ as a first approximation to θ, apply the Newton-Raphson procedure once to the function $f(\theta) = 2\sin\theta - \theta$ to find a second approximation to θ. Give your answer to two decimal places.

c i Solve the equation $2\cos\theta = 1$ for $0 \leqslant \theta \leqslant \frac{1}{2}\pi$. Give your answer in terms of π.

ii Starting with $\theta_1 = 1$, calculate the value of the iterate θ_2, produced by the Newton-Raphson procedure when applied to $f(\theta)$.

iii Explain why θ_2 is significantly different from θ_1. Justify your answer.

Review 3

1 Let $f(x) = 2x^2 + \frac{1}{x} - 5$, where $x > 0$.

a Show that the equation $f(x) = 0$ has a root, α, in the interval (1.4, 1.6).

b Starting with the interval (1.4, 1.6) use interval bisection twice to find an interval of width 0.05 which contains α.

c Hence write down α correct to one decimal place.

2 The diagram shows an accurate sketch of the graph of the equation $y = \sqrt{x} - 3^{x-1} + 7$ for $x \geqslant 0$. The graph crosses the x-axis at α, where $2 < \alpha < 3$.

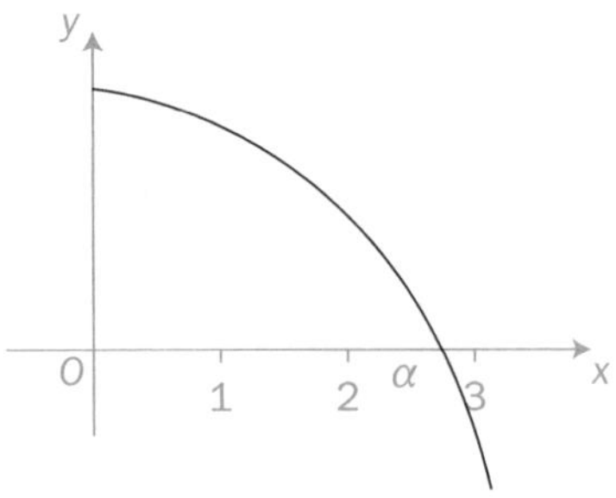

a Use linear interpolation on the end points of the interval (2, 3) to find an approximation to α. Give your answer to two decimal places.

b Explain, with the aid of the diagram, why the answer to part **a** is an under-estimate for α.

c Hence write down α correct to one decimal place.

3 It is given that the equation $2x^3 - 5x - 5 = 0$ has exactly one positive root, α.

a Show that $1 < \alpha < 2$.

b Taking $x_1 = 2$ as a first approximation to α, apply the Newton-Raphson process twice to $f(x)$ to find a second approximation to α.

c Show that as an approximation to α the value of x_2 found in part **b** is accurate to two decimal places.

4 The diagram shows part of the graph $y = f(x)$ where $f(x) = x\cos(3x - 1) - 3$. The graph crosses the x-axis at α and β, where $\alpha < \beta$.

It is given that α lies in the interval (4, 4.4).

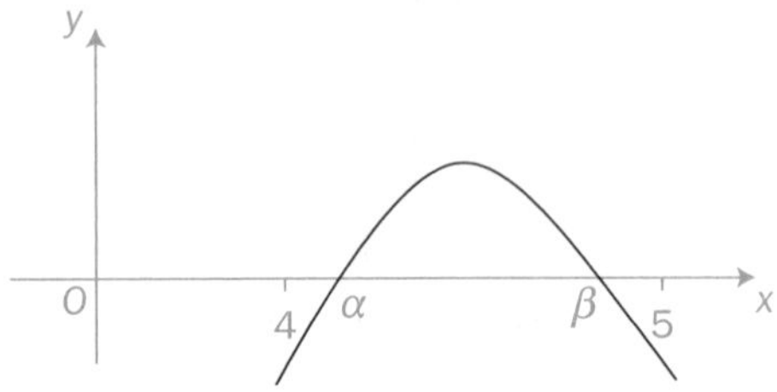

a Starting with the interval (4, 4.4), use interval bisection three times to find α correct to one decimal place.

b Show that β lies in the interval (4.6, 5).

c Briefly explain why applying linear interpolation once to the end points of the interval (4, 5) will not produce a reliable approximation for either of these roots.

FP1

5 It is given that the equation $f(x) = 0$, where $f(x) = 3\sqrt[3]{x} - 4x + 10$, has exactly one root, α.

a Show that α lies in the interval (3, 4).

b Using the end points of the interval (3, 4), find by linear interpolation an approximation to α. You should give intermediate values and your final answer to at least two decimal places.

c Taking $x_1 = 4$ as a first approximation to α, apply the Newton-Raphson process once to $f(x)$ to find a second approximation to α. Give your answer to two decimal places.

d Show that, as an approximation to α, exactly one of the answers to parts **b** and **c** is accurate to two decimal places.

6 The diagram shows the graph with equation $y = f(\theta)$, where $f(\theta) = \sin(2\theta) - \theta$. The graph crosses the positive θ-axis exactly once, at α.

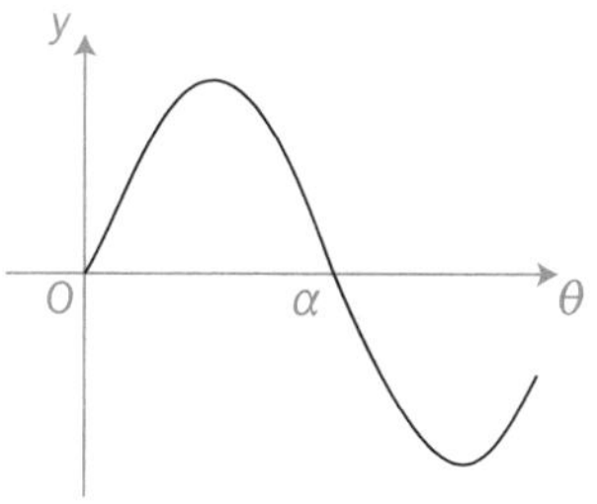

It is given that $0.8 < \alpha < 1$.

a Starting with the interval (0.8, 1) use interval bisection twice to find an interval of width 0.05 which contains α. Hence write down α to one decimal place.

b Using the end points of the interval found in part **a** find by linear interpolation an approximation to α. Show that this estimate is accurate to two decimal places.

7 The diagram shows the curve with equation $y = f(x)$ where

$$f(x) = 2x^3 - x^2 - 9x + 1$$

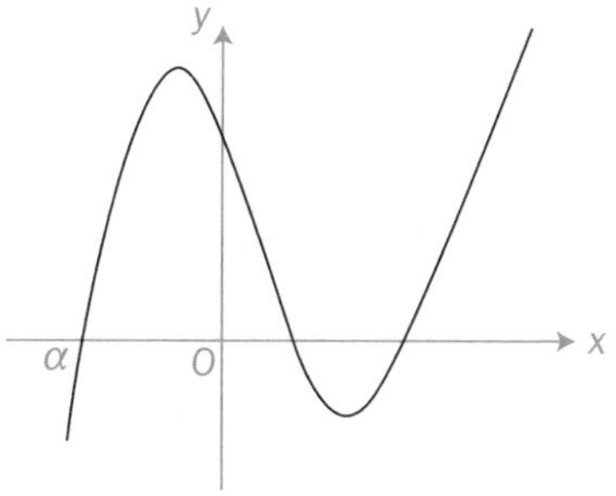

The graph crosses the negative x-axis at α only.

a Show that $-2 < \alpha < -1$.

b Taking $x_1 = -2$ as a first approximation to α, apply the Newton-Raphson process once to $f(x)$ to find a second approximation to α. Show that your estimate is accurate to two decimal places.

c Taking $x_1 = -1$ as a first approximation to α, apply the Newton-Raphson process once to $f(x)$ to find another approximation to α.
Explain why your answer is such a poor approximation to α. Justify your answer.

8 It is given that the equation $f(x) = 0$,

where $f(x) = \frac{2\sqrt{x} - 1}{x} - \frac{x^2}{2} + 2$, for $x > 0$,

has exactly one root α in the interval $[2, 3]$.

a Using the end points of the interval $[2, 3]$, find by linear interpolation an approximation to α.
Give your answer to three significant figures.

b Taking your answer to part **a** as a first approximation to α, use the Newton-Raphson procedure once on $f(x)$ to find a second approximation, x_2, to α.

c Show that, as an approximation to α, the value of x_2 found in part **b** is accurate to four significant figures.

9 $f(x) = \pi\sin(2x) - 2x$ where x is in radians.

a Show by drawing a sketch that the equation $f(x) = 0$ has exactly one positive root, α, where $\frac{1}{4}\pi < \alpha < \frac{1}{2}\pi$.

b Using the end points of the interval $\left(\frac{1}{4}\pi, \frac{1}{2}\pi\right)$, find by linear interpolation an approximation to α.
Give your answer in terms of π.

c Determine whether the approximation to α found in part **b** is an under-estimate or an over-estimate.

10 The equation $f(x) = 0$, where $f(x) = (x - 2\sqrt{x})^2 - 2$, $x > 0$, has exactly one root, α.

a Show that α lies between $x = 6.5$ and $x = 6.6$

b Using the end points of the interval $(6.5, 6.6)$, find by linear interpolation an approximation to α.
Give your answer to two decimal places.

c Show that taking $x_1 = 4$ as a starting value when applying the Newton-Raphson process to $f(x)$ to find an approximation to α leads to an error.
Explain the cause of this error.

FP1

11 $f(x) = x^2 + 1 - \frac{3}{x}, x \neq 0$

a By sketching on the same diagram the graphs of $y = x^2 + 1$ and $y = \frac{3}{x}$, show that the equation $f(x) = 0$ has exactly one real root α.

b Given that α lies in the interval [1.20, 1.22] use interval bisection twice to find an interval of width 0.005 which contains α.
State the value of α to 2 decimal places.

c Taking $x_1 = 1.2$ as a first approximation to α, apply the Newton-Raphson process once to $f(x)$ to find a second approximation x_2 to α.

d Show that, as an approximation to α, x_2 is accurate to four decimal places.

12

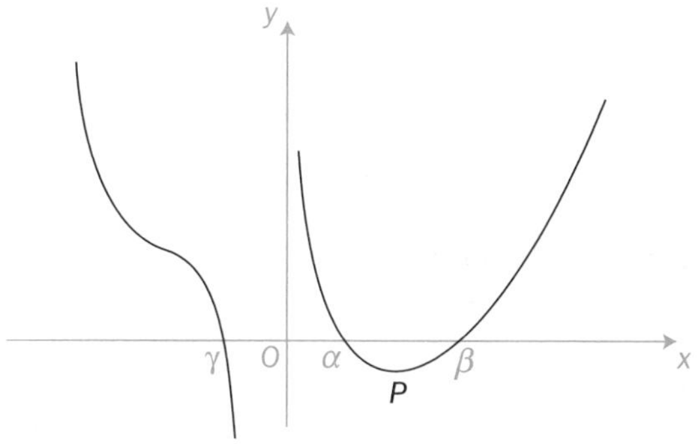

The diagram shows the curve with equation $y = f(x)$ where $f(x) = \frac{2}{x} + \frac{1}{2}x^2 - 4, x \neq 0$.

The graph has roots α, β and γ, and a stationary point at P. It is given that, when applied to $f(x)$ with a starting value of $x_1 = 1$, the Newton-Raphson process converges to one of these three roots.

a Use this starting value to calculate the value of x_2 and hence determine the root to which further iterations would converge.

b Find the x-coordinate of point P.

c Hence, without further calculation, state the root to which the Newton-Raphson process converges when applied to $f(x)$ with a starting value of $x_1 = 1.5$

3

Exit

Summary

Refer to

- If a continuous function f(x) changes sign across an interval (a, b) then the equation f(x) = 0 has a root, α, between $x = a$ and $x = b$ 3.1
- You can find a more accurate estimate for a root by bisecting an interval which is known to contain that root. 3.2
- You can estimate a root of an equation f(x) = 0 by assuming that the graph of $y =$ f(x) is a straight line over an interval which contains the root. 3.3
- If the equation f(x) = 0 has a root, $x = \alpha$, and $x = x_1$ is a first approximation to α, then the iterates $x_2, x_3, x_4, \ldots$ produced by the Newton–Raphson formula $x_{n+1} = x_n - \frac{f(x_n)}{f'(x_n)}$ will generally be increasingly accurate approximations to α. 3.4
- Starting near a stationary point can cause the Newton-Raphson process to become unstable. 3.4

Links

Numerical methods were developed to provide appoximate solutions to equations that were difficult (or impossible) to solve. Nowadays the Newton-Raphson procedure is used by computers to solve much harder problems than quadratic equations.

An increase in the number of applied mathematicians working in the world of finance has led to mathematical models and numerical techniques being used to solve problems in this field such as portfolio management and derivatives pricing.

The study of how the error produced by a numerical method behaves is called numerical analysis. This is an important tool for estimating the degree of confidence in the reliability of any approximation found using a numerical method.

4 Coordinate systems

This chapter will show you how to
- sketch a parabola or rectangular hyperbola
- represent a parabola as a locus of points
- describe points on a parabola or rectangular hyperbola using a parameter
- find an equation for the tangent or normal at any point on either of these curves.

Before you start

You should know how to:

1 Find the coordinates of the point(s) at which two curves intersect.

e.g. Find the coordinates of the points where the line $y = 4 - x$ intersects the circle $x^2 + y^2 = 10$

$y = 4 - x$ so $y^2 = (4 - x)^2 = 16 - 8x + x^2$

Substitute for y^2 in $x^2 + y^2 = 10$:

$x^2 + (16 - 8x + x^2) = 10$

so $x^2 - 4x + 3 = 0$

i.e. $(x - 1)(x - 3) = 0$,

so either $x = 1$ or $x = 3$

Use $y = 4 - x$ to find the corresponding y-coordinates:

The line and circle intersect at (1, 3) and (3, 1).

2 Find an equation for a tangent or normal to a curve.

e.g. Find an equation for the tangent to the curve $y = x^2 + 2x^{-1}$ at the point $P(2, 5)$.

Differentiate: $\frac{dy}{dx} = 2x - 2x^{-2}$

When $x = 2$, $\frac{dy}{dx} = 4 - 0.5 = 3.5$

The equation of the tangent at P is $y - 5 = 3.5(x - 2)$

$\Rightarrow \quad y = \frac{7}{2}x - 2$

Check in:

1 Find the point(s) at which each pair of curves intersect.

See C1 for revision. Refer to Section 1.1.

a $y = x^2 - 4x + 1$ and $y = 2x - 7$

b $y = 3x - 2$ and $y = \frac{1}{x}$

c $y = x^2 + 3x + 1$ and $y = \frac{1}{x + 1}$

d $y = x + 5$ and $x^2 + y^2 = 13$

2 Find equations for the tangent and normal to the curve C at the point P.

See C1 for revision. Refer to Section 1.5.

a C: $y = x^3 + 2x^2 - 1$ at $P(1, 2)$

b C: $y = x + 2\sqrt{x}$ at $P(4, 8)$

c C: $y = 4x - \frac{1}{x}$ at $P(-1, -3)$

4.1 The parabola

A parabola has an equation of the form $y^2 = 4ax$, where a is a positive constant.

You can sketch the graph of a parabola by plotting particular points on the curve.

EXAMPLE 1

Sketch the parabola with equation $y^2 = 4x$

Choose appropriate values of x:

Since $y^2 \geqslant 0$ and $a > 0$, the curve is not defined for negative values of x.

When $x = 0$, $y^2 = 4 \times 0 = 0$, so $y = 0$
The curve passes through the origin (0, 0)

There are two possible points on the curve for each value of $x > 0$.

When $x = 1$, $y^2 = 4 \times 1 = 4$, so $y = \pm 2$
The curve passes through the points (1, 2) and (1, −2)

When $x = 4$, $y^2 = 4 \times 4 = 16$, so $y = \pm 4$
The curve passes through the points (4, 4) and (4, −4)

When $x = 9$, $y^2 = 4 \times 9 = 36$, so $y = \pm 6$
The curve passes through the points (9, 6) and (9, −6)

Plot these points and join them with a smooth curve:

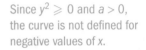

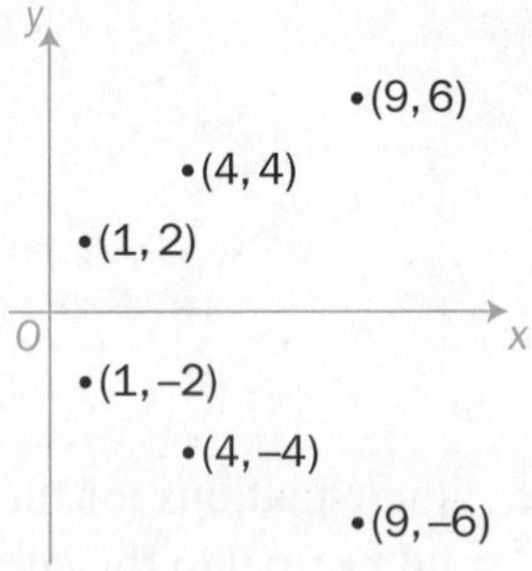

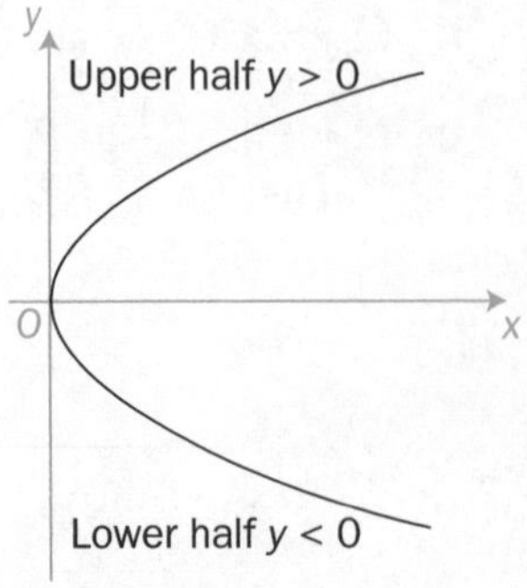

FP1

The general parabola has equation $y^2 = 4ax$ where $a > 0$.
Given $a > 0$, you can define a fixed point $F(a, 0)$ on the positive x-axis and a fixed vertical straight line L with equation $x = -a$.
Point F is the focus and the line L is the directrix of the parabola.

The focus and directrix have important geometrical properties.

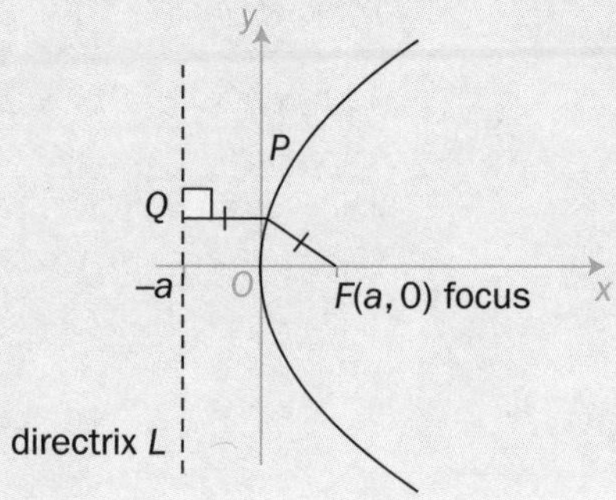

The directrix has equation $x = -a$

EXAMPLE 2

A parabola C has equation $y^2 = 6x$

a Find the coordinates of the focus of C and the equation of its directrix.

b Find the points where the line with equation $y = 2x - 6$ intersects C.

---

a Compare $y^2 = 6x$ with $y^2 = 4ax$:

$$4a = 6, \text{ so } a = \frac{3}{2}$$

Hence the focus has coordinates $\left(\frac{3}{2}, 0\right)$ and the directrix has equation $x = -\frac{3}{2}$

b Solve the simultaneous equations:

$$y^2 = 6x \quad (1)$$
$$y = 2x - 6 \quad (2)$$

Substitute $y = 2x - 6$ in (1): $(2x - 6)^2 = 6x$

$$4x^2 - 24x + 36 = 6x$$
$$2x^2 - 15x + 18 = 0$$
$$(2x - 3)(x - 6) = 0$$
$$\text{so } x = \frac{3}{2}, x = 6$$

Rearrange and divide through by 2.

Substitute for x into equation (2) to find y:

Eqn (2) is easier to work with than eqn (1).

When $x = \frac{3}{2}$, $y = -3$
When $x = 6$, $y = 6$

Hence $y = 2x - 6$ intersects C at the points $\left(\frac{3}{2}, -3\right)$ and $(6, 6)$.

FP1

If C is a parabola with equation $y^2 = 4ax$ then any point P on C is equidistant from its focus $F(a, 0)$ and its directrix $x = -a$. This is known as the focus-directrix property of a parabola.

$FP = PQ$ in the diagram.

You can establish the focus-directrix property by using the equation of the parabola.

In the diagram, $P(x, y)$ is any point on the parabola and $F(a, 0)$ is the focus.
Point Q on L is such that PQ is the (perpendicular) distance of P from L.

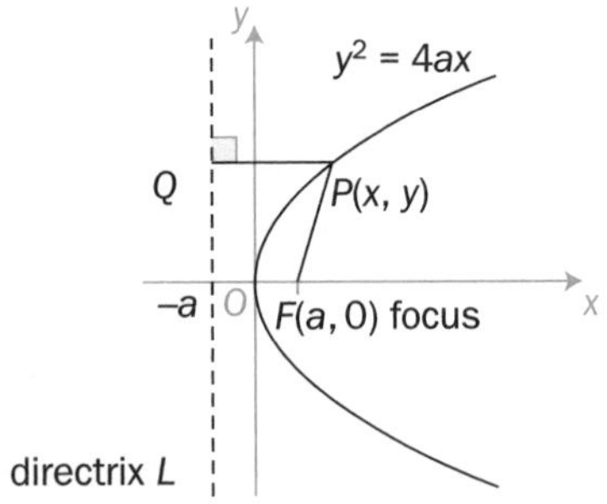

Since the line PQ is horizontal
$PQ = x + a$

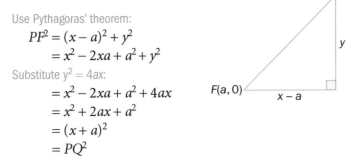

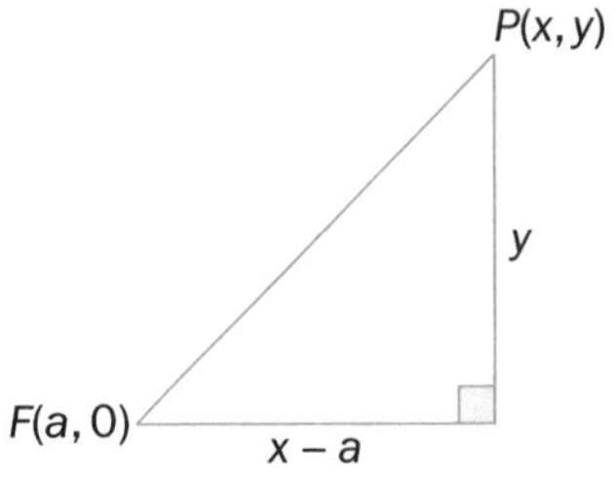

The distance between points $P(a, b)$ and $Q(c, d)$ is given by $PQ = \sqrt{(a-c)^2 + (b-d)^2}$

Use Pythagoras' theorem:

$$PF^2 = (x - a)^2 + y^2$$
$$= x^2 - 2xa + a^2 + y^2$$

Substitute $y^2 = 4ax$:

$$= x^2 - 2xa + a^2 + 4ax$$
$$= x^2 + 2ax + a^2$$
$$= (x + a)^2$$
$$= PQ^2$$

This proves that, for any point P on a parabola, $PF = PQ$

You can therefore define a parabola as the locus of points equidistant from a fixed point and a fixed straight line.
You can solve geometrical problems using the focus-directrix property of a parabola.

A locus is a set of points satisfying a given condition.

EXAMPLE 3

The diagram shows a parabola W, with focus F and directrix L. Point P on W is such that the line FP makes an angle of 80° with the positive x-axis. Point Q on L is such that the line PQ is parallel to the x-axis.

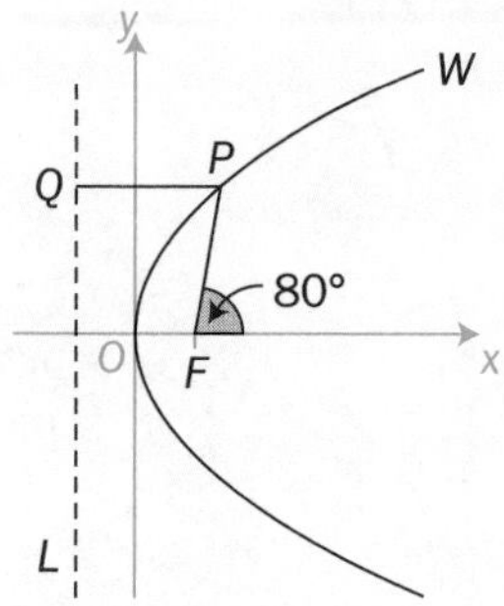

Find angle PFQ.

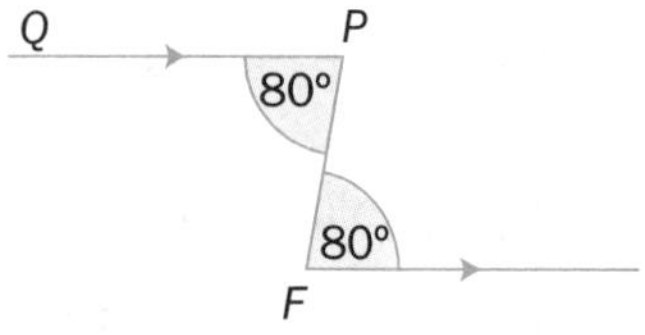

Since the line PQ is parallel to the x-axis,

- PQ is the perpendicular distance of P from the directrix L
- angle $QPF = 80°$ (alternate angles).

By the focus-directrix property, $PF = PQ$ and hence triangle QPF is isosceles.

$$\theta = \frac{1}{2}(180° - 80°)$$
$$= 50°$$

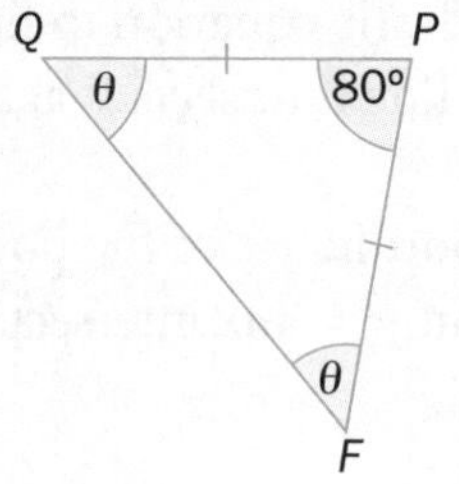

Hence angle $PFQ = 50°$

Exercise 4.1

1 Write down the coordinates of the focus, F, and the equation of the directrix, L, for each parabola.

a $y^2 = 12x$ **b** $y^2 = 10x$ **c** $y^2 = \frac{1}{2}x$ **d** $y^2 = \sqrt{48}x$

2 **a** Use algebra to show that the curves with equations $y^2 = 9x$ and $y^2 = 3x$ intersect only at the origin.

b Find the horizontal distance between the foci of these parabolas.

The plural of *focus* is *foci*.

c Find the x-coordinate of the point on each curve where $y = 3\sqrt{2}$

d Sketch the curves on the same diagram. Clearly label each curve with its equation.

FP1

3 A parabola, C, is defined by the equation $y^2 = kx$, where k is a constant.

a Given that C passes through the point $P(4, 6)$, find the value of k.

b Hence find the coordinates of the focus, F, and the equation of the directrix, L, of C.

4 a On the same diagram sketch the parabola with equation $y^2 = 3x$ and the line with equation $y = 2x$

b Find the coordinates of the point P, not at the origin, where this line and curve intersect.

c Find the area of triangle OFP where F is the focus of the parabola.

5 A parabola, V, has equation $y^2 = 8x$

a Find the value of k for which the point $P(k, 2)$ lies on V.

b Show that the line passing through point P and the focus of V has equation given by $4x + 3y - 8 = 0$

c Solve an appropriate quadratic equation to find the coordinates of the point Q where this line intersects V again.

6 a Find, in terms of a, the coordinates of the point P where the parabola with equation $y^2 = 4ax$ intersects the line $y = a$ for $a > 0$.

b Show that the equation of the line passing through point P and the focus F of this parabola is given by the equation $3y + 4x = 4a$

c Find, in terms of a, the coordinates of the point Q where the line in part **b** intersects the directrix of the parabola.

7 A parabola C has equation $y^2 = 20x$

a Find the coordinates of the focus F of C.

Point P on C lies 13 units from point F.

b Using the focus-directrix property, or otherwise, find the x-coordinate of P.

c Hence find the two possible y-coordinates of P. Give your answers in simplified surd form.

d Show that the area of triangle OFP is $10\sqrt{10}$.

8 The diagram shows a parabola, C, with equation $y^2 = 4x$, focus F and directrix L.

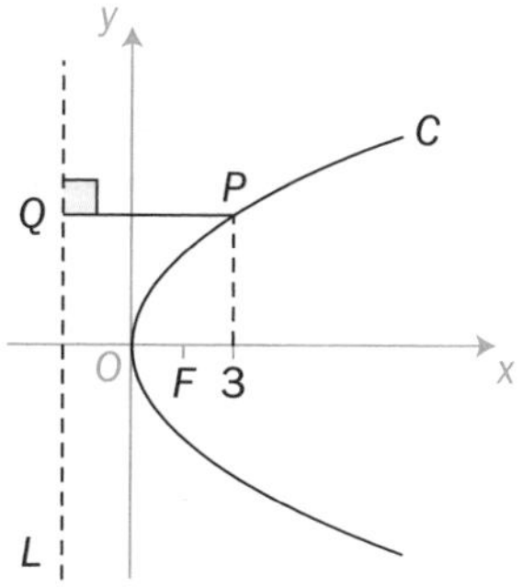

Point P on the upper-half of C has x-coordinate 3.
Point Q on L is such that the line PQ is horizontal.

a Show that the y-coordinate of P is $2\sqrt{3}$

b Find the coordinates of Q.

c Find the coordinates of F and show that the line FP makes an angle of 60° with the positive x-axis.

d Hence, or otherwise, completely describe the triangle FPQ and find its exact area.

FP1

9 Taking the definition of a parabola, C, as the locus of points equidistant from a fixed point, $F(a, 0)$, and a fixed line with equation $x = -a$, where $a > 0$, show that

a C passes through the origin

b every point on C lies at least a units from F

c C is symmetrical in the x-axis.

10 A parabola, D, has equation $y^2 = 6x$
A circle, E, has centre $O = (0, 0)$ and radius 4.

a Sketch on the same diagram the graphs of D and E. Show clearly the focus, F, and directrix, L, of D.

b Find the exact coordinates of the points where D and E intersect.

c Using the cosine rule, or otherwise, show that $\cos OFP = -\frac{1}{7}$ where P is a point of intersection of D and E.

4.2 The rectangular hyperbola

A **rectangular hyperbola** is a curve with equation $xy = c^2$ for $c > 0$, a constant.

A rectangular hyperbola is not defined for $x = 0$ or $y = 0$.

You can write this equation as $y = \frac{c^2}{x}$, $x \neq 0$

You can sketch the complete graph of a rectangular hyperbola by plotting particular points on the curve for $x > 0$ and then rotating this section 180° about the origin, O.

EXAMPLE 1

Complete the table and hence sketch the rectangular hyperbola with equation $xy = 6$

x	1	2	3	4
y				

In this example $c^2 = 6$ and so $c = \sqrt{6}$

Make y the subject of the equation:

$$y = \frac{6}{x}$$

Calculate the value of y for each given value of x:

x	1	2	3	4
$y = \frac{6}{x}$	6	3	2	1.5

e.g. When $x = 2$, $y = \frac{6}{2} = 3$

Plot the points and join them to make a smooth curve and then use this to draw the complete curve:

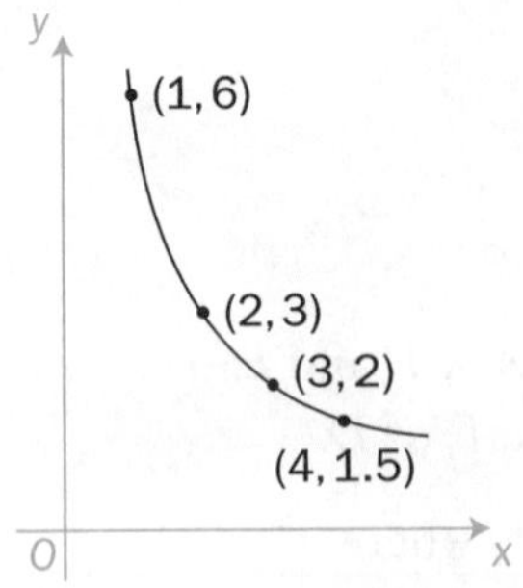

The axes are asymptotes to the curve.

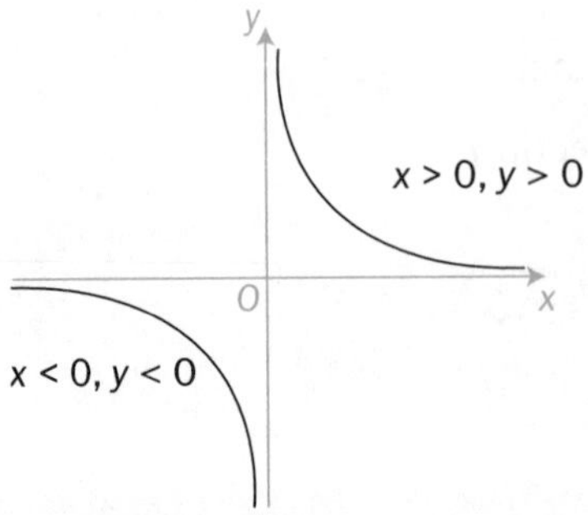

EXAMPLE 2

a Sketch on the same diagram the rectangular hyperbola with equation $xy = 16$ and the line with equation $y = 3x - 2$

b Use algebra to find the coordinates of the points where the line and curve intersect.

---

a

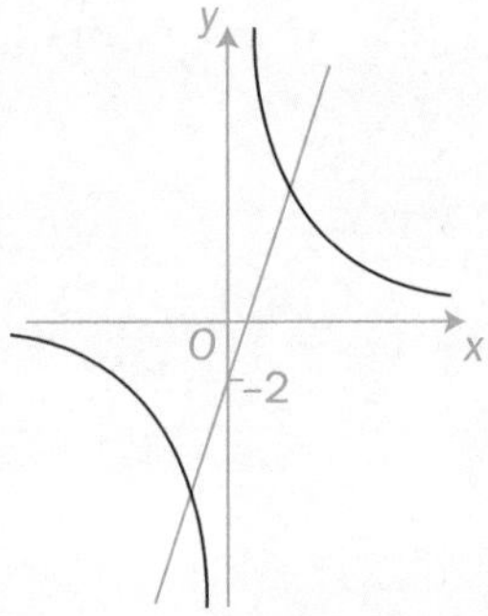

All rectangular hyperbolae have the same appearance.

b Solve the simultaneous equations:

$$xy = 16 \quad (1)$$
$$y = 3x - 2 \quad (2)$$

Substitute $3x - 2$ for y in (1):

$$x(3x - 2) = 16$$

Expand the bracket and rearrange.

$$3x^2 - 2x - 16 = 0$$
$$(3x - 8)(x + 2) = 0$$

Factorise.

$$x = \frac{8}{3}, x = -2$$

Substitute for x in (2) to find y:

Eqn (2) is easier to work with than eqn (1).

if $x = \frac{8}{3}, y = 3 \times \frac{8}{3} - 2 = 6$

if $x = -2, y = 3 \times (-2) - 2 = -8$

Hence the points of intersection are $\left(\frac{8}{3}, 6\right)$ and $(-2, -8)$.

Exercise 4.2

1 **a** Sketch on the same diagram the rectangular hyperbola with equation $xy = 9$ and the line with equation $y = 4x$

b Find the coordinates of the points where this line and curve intersect.

2 A rectangular hyperbola has equation $xy = c^2$, for c a positive constant. Which of these points lie on this curve?

a $(c, -c)$ **b** $\left(\sqrt{3c^2}, \frac{\sqrt{3}}{3}c\right)$

c $(c + 1, c - 1)$ **d** $\left(1 + \frac{1}{c}, \frac{c^3}{c + 1}\right)$

FP1

3 A rectangular hyperbola has equation $xy = c^2$, for c a positive constant. The curve passes through the point $\left(-2\sqrt{3}, -\sqrt{3}\right)$.

a Find the value of c.

b Find the value of k for which the point $P\left(-\frac{3}{4}, k\right)$ lies on this curve.

4 Find the coordinates of the point(s) of intersection of

a the rectangular hyperbola $xy = 36$ and the line $y = 2x + 1$

b the rectangular hyperbola $xy = 16$ and the curve $y = \frac{1}{4}x^2$

c the rectangular hyperbola $xy = 7$ and the cubic curve $y = x^3 + 6x$

5 For any rectangular hyperbola C with equation $xy = c^2$, for c a positive constant, use algebra to

a show that any straight line with negative gradient which passes through the origin does not intersect C

b determine whether or not the line with equation $y = 2c - x$ intersects C.

6 The diagram (not to scale) shows part of a rectangular hyperbola H with equation $xy = c^2$, for $c > 0$. The line L with equation $2y + 3x = 15$ intersects the curve at points $P(2, 4.5)$ and Q and the x-axis at point R.

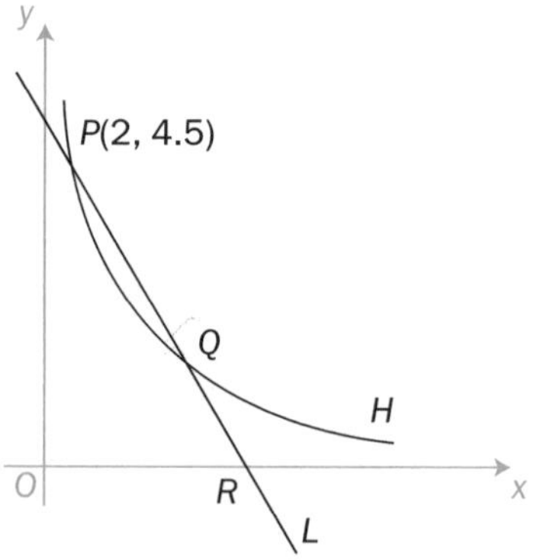

a Show that $c = 3$.

b Find the coordinates of point Q.

c Find the coordinates of point R and hence show that the area of triangle OQR is twice the area of triangle OQP.

7 The diagram shows the graph of the rectangular hyperbola H with equation $xy = 12$

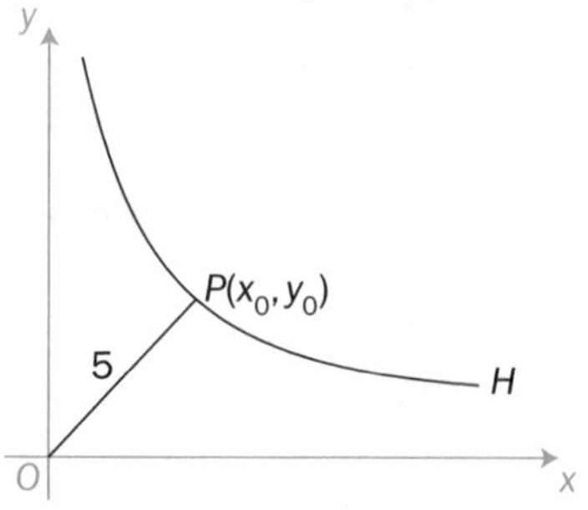

Point $P(x_0, y_0)$ on H, where $x_0 > 0$, is such that the line OP has length 5.

a Use the coordinates of P to show that $x_0^2 + y_0^2 = 25$

b Deduce that $x_0^4 - 25x_0^2 + 144 = 0$ and solve this equation to find the possible coordinates of point P.

8 The diagram shows a circle C defined by the equation $x^2 + y^2 = 9$ and a rectangular hyperbola D with equation $xy = \sqrt{8}$. The two curves intersect at points P, Q, R and S.

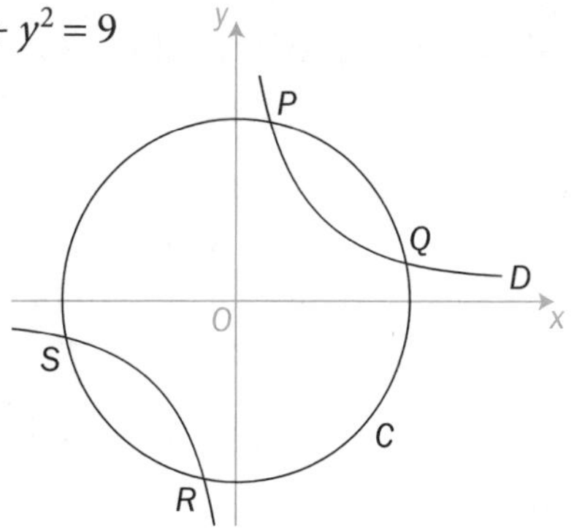

a Find the coordinates of P, Q, R and S.

b Find the equation of the line PR.

c Hence show that the line PR is a diameter of this circle and state the value of angle RSP.

9 The diagram shows part of the graph of the rectangular hyperbola C with equation $xy = c^2$

Point $P(c, c)$ lies on C.

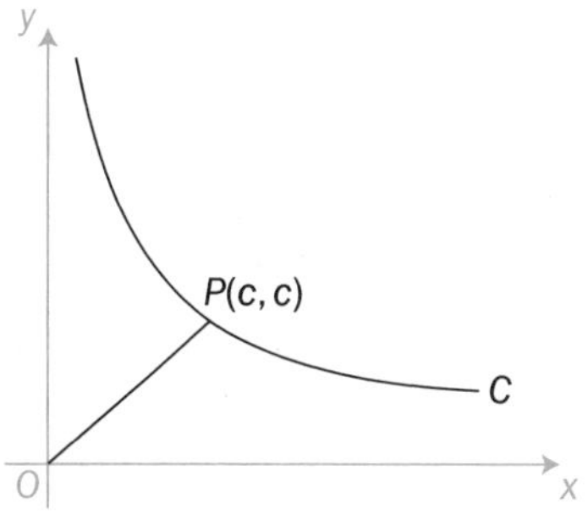

a Show that $OP = \sqrt{2}c$

Let $Q(x, y)$ be any point on C, where $x > 0$.

b By expanding $(x - y)^2$, or otherwise, show that P is the point on C which is closest to the origin.

10 The diagram shows part of a rectangular hyperbola H with equation $xy = c^2$ where $c > 0$. Points P, Q and R lie on H. The x-coordinates of points P, Q and R are $\frac{1}{2}c$, c and $\frac{3}{2}c$, respectively.

a Find, in terms of c, the y-coordinates of each of these points.

b By using the trapezium rule with two strips, show that the area A between the curve $xy = c^2$ and the x-axis from $x = \frac{1}{2}c$ to $x = \frac{3}{2}c$ is less than $\frac{7}{6}c^2$.

c Show further that $5c^2 < 6A < 7c^2$

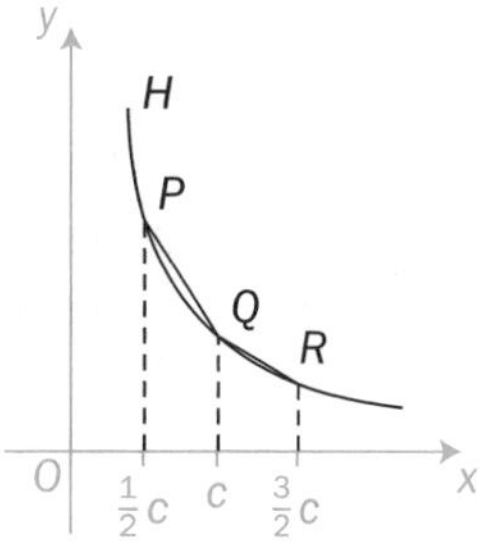

11 The diagram shows part of a rectangular hyperbola with equation $xy = c^2$, where c is a positive constant.

$P(x_0, y_0)$ is any point on the curve and points A and B on the x-and y-axes, respectively, are such that $OAPB$ is a rectangle.

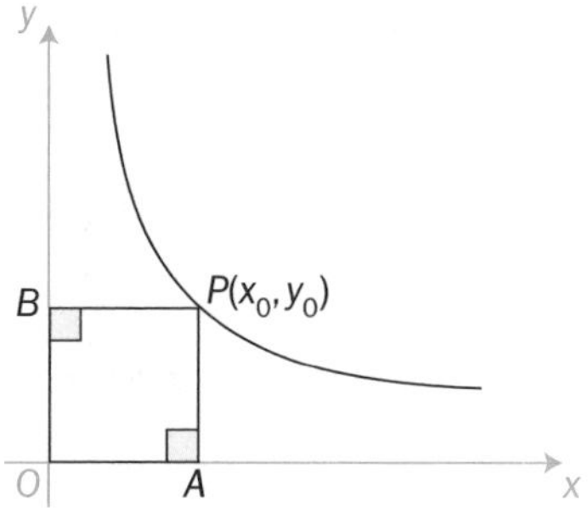

a Given that the area of OAPB is 8, find the exact value of c.

b Show that $T^2 = 4(OP^2 + 16)$, where T is the perimeter of OAPB.

A cartesian equation of a curve involves only the variables x and y.
For example, $y^2 = 4ax$ is the cartesian equation of a parabola.
You can express the equation of this parabola by defining x and y in terms of a third variable, t (called a parameter).

For each point $P(x, y)$ on the parabola it is convenient to define t as $\frac{y}{2a}$.

The definition makes sense as $a \neq 0$.

Rearrange to make y the subject:

$$t = \frac{y}{2a}$$
$$y = 2at$$

Substitute $y = 2at$ into the equation of the parabola:

$$y^2 = 4ax$$
$$(2at)^2 = 4ax$$
$$4a^2t^2 = 4ax$$
$$x = at^2$$

$4a^2t^2 = 4ax$, so $x = \frac{4a^2t^2}{4a} = at^2$

Hence for any value of the parameter t the point $P(at^2, 2at)$ lies on the parabola with equation $y^2 = 4ax$

Each value of t produces exactly one point $P(at^2, 2at)$ on the parabola.

Every point P on the parabola with cartesian equation $y^2 = 4ax$ can be expressed in parametric form $P(at^2, 2at)$, where $t \in \mathbb{R}$

$t \in \mathbb{R}$ means t takes all possible real values.

EXAMPLE 1

A parabola has cartesian equation $y^2 = 8x$
Find the coordinates of the point on the parabola corresponding to the value $t = \frac{3}{2}$

Comparing $y^2 = 4ax$ with $y^2 = 8x$ gives $a = 2$

If $4a = 8$, $a = 2$

The point corresponding to the value $t = \frac{3}{2}$ has coordinates $(at^2, 2at)$

When $t = \frac{3}{2}$, $x = at^2$ and $y = 2at$

$$x = 2 \times \left(\frac{3}{2}\right)^2 = \frac{9}{2} \qquad y = 2 \times 2 \times \frac{3}{2} = 6$$

Hence the point has coordinates $\left(\frac{9}{2}, 6\right)$.

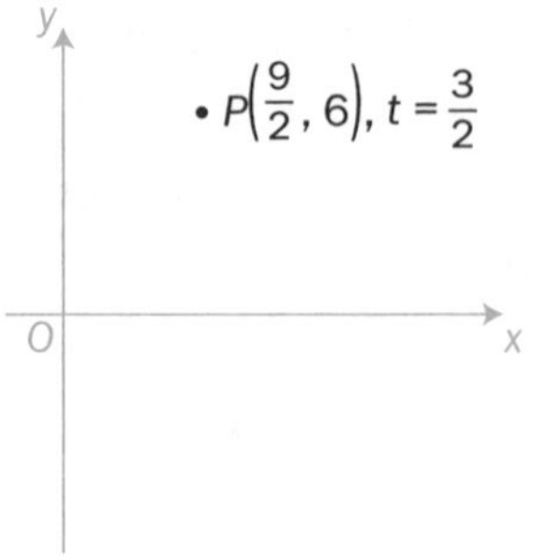

Every point, P, on a rectangular hyperbola with equation $xy = c^2$ can be expressed in parametric form $P\left(ct, \frac{c}{t}\right)$, $t \in \mathbb{R}$ $t \neq 0$

The curve is not defined when $t = 0$

EXAMPLE 2

Show that any point $\left(3t, \frac{3}{t}\right)$, where $t \neq 0$, lies on the rectangular hyperbola with cartesian equation $xy = 9$

For the point $\left(3t, \frac{3}{t}\right)$, $x = 3t$ and $y = \frac{3}{t}$

Evaluate xy:

$$xy = 3t \times \frac{3}{t}$$

$$= \frac{9t}{t} = 9, \text{ as required}$$

Exercise 4.3

1 Express in parametric form the coordinates of the point P which lies on

a the rectangular hyperbola with cartesian equation $xy = 16$

b the parabola with cartesian equation $y^2 = 6x$

c the rectangular hyperbola with cartesian equation $xy = \frac{1}{9}$

d the parabola with cartesian equation $y^2 = \sqrt{32}x$

2 a Write down, in terms of t, the coordinates of any point on the parabola C with cartesian equation $y^2 = 12x$

b In each column of the table, the parameter t corresponds to a point $P(x, y)$ on the upper half of this parabola.

Copy and complete the table.
Give values in simplified surd form where appropriate.

t	0				2	
x			$\frac{1}{3}$			1
y		3		$3\sqrt{2}$		

3 The parabola, C, with equation $y^2 = 5x$ passes through the point $P(k, -10)$

a Show that $k = 20$

b Write down, in parametric form, the coordinates of any point on C.

c Find the value of t corresponding to point P.

FP1

4 The coordinates of any point, P, on a rectangular hyperbola, H, can be expressed in the form $\left(\sqrt{18}t, \frac{\sqrt{18}}{t}\right)$, $t \neq 0$

a Find the coordinates of the point on H corresponding to $t = \sqrt{8}$

b Find the value of t corresponding to the point on H with y-coordinate $-\sqrt{3}$.

Give your answer in the form $a\sqrt{b}$ for integers a and b to be stated.

c Write down the cartesian equation of H.

5 It is given that, in parametric form, the coordinates of any point, P, on a parabola, C, are $P(3t^2, 6t)$

a Sketch this parabola, clearly marking the position of its focus, F, and its directrix.

b Find the exact two values of t corresponding to a point P on C such that $FP = 3 \times OF$

6 A rectangular hyperbola, R, has cartesian equation, $xy = \frac{1}{4}$

a Write down, in parametric form, the coordinates of any point on R.

b Find the equation of the line which passes through points P and Q on R corresponding to the values $t = \frac{2}{3}$ and $t = -\frac{3}{2}$ respectively.

7 The diagram shows the rectangular hyperbola C with cartesian equation $xy = c^2$, where c is a constant.

$P(x_0, y_0)$ is a point on C corresponding to the value t_0.

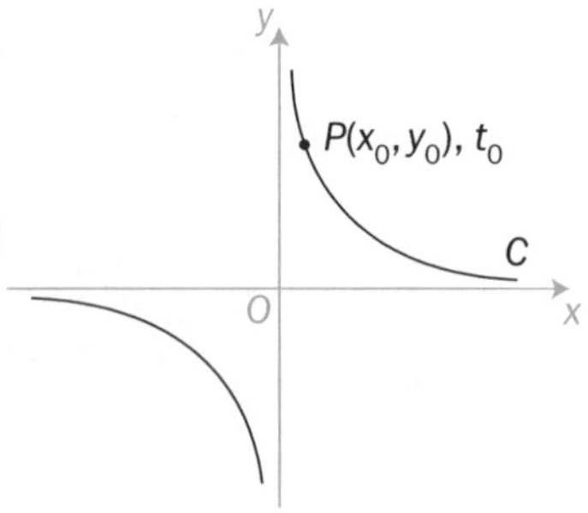

a Express the coordinates of P in parametric form.

b Find, in terms of x_0 and y_0, the coordinates of the point on C corresponding to the value

i $t = -t_0$ ii $t = \frac{1}{t_0}$

8 The coordinates of any point on a rectangular hyperbola H are given by $\left(ct, \frac{c}{t}\right)$ for any real number $t \neq 0$, where c is a positive constant. The line L with equation $y = 2x + c$ intersects H at the points P and Q as shown in the diagram.

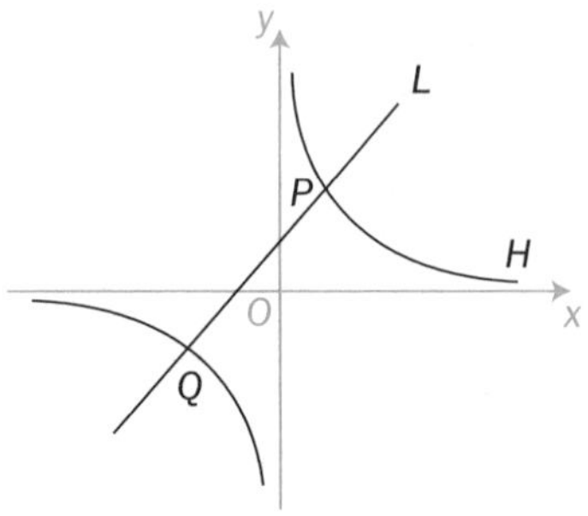

a Show that if $\left(ct, \frac{c}{t}\right)$ are the coordinates of any point of intersection of L and H then $2t^2 + t - 1 = 0$

b Solve this quadratic equation and hence find, in terms of c, the coordinates of the points P and Q.

FP1

9 Every point on a parabola, G, has coordinates given by $(4t^2, 8t)$ for $t \in \mathbb{R}$.
The line l has cartesian equation $3y + 4x = 16$

a Show that any value of t corresponding to a point of intersection of l and G satisfies the equation
$2t^2 + 3t - 2 = 0$

b Solve this equation and hence find the coordinates of the points P and Q where l and G intersect.

c Verify that l passes through the focus, F, of G.

d Use the focus-directrix property to show that the distance $PQ = 25$

10 The diagram shows part of a parabola C with focus F. Any point P on C has coordinates given by $(t^2, 2t)$ where t is any real number.

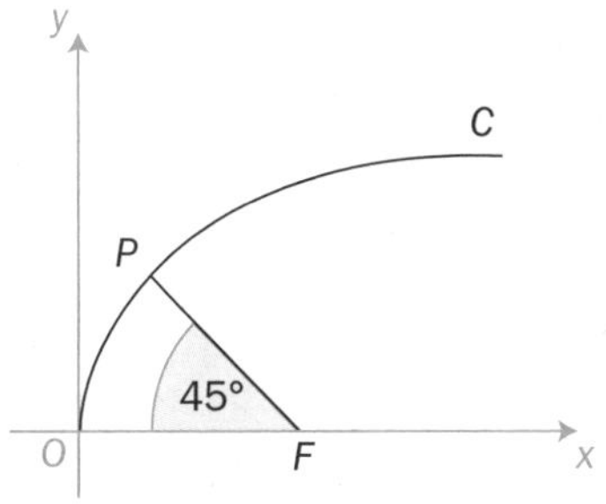

a Write down the coordinates of F.

Point $P(t^2, 2t)$ on C is such that angle $PFO = 45°$

b Using the focus-directrix property of C, or otherwise, show that $PF = 1 + t^2$

c Find an expression, in terms of t, for OP^2 and hence show that $t = \sqrt{2} - 1$

d Find the exact area of triangle OPF, giving your answer in the form $\sqrt{p} + q$ for integers p and q to be stated.

11 The diagram shows part of a rectangular hyperbola, C.
$P\left(\sqrt{2}t, \frac{\sqrt{2}}{t}\right)$ is any point on C.

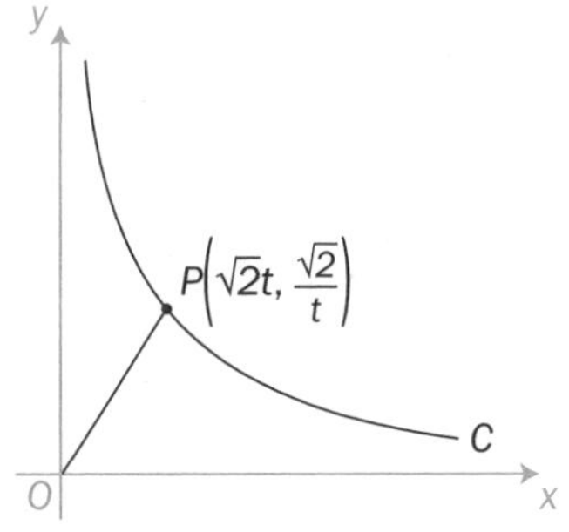

a Show that the gradient of the line OP is given by $\frac{1}{t^2}$

Point Q on the line $y = k - 2x$, where $k > 0$ is a constant, is such that the distance OQ is as short as possible.

b Given that Q is also on C, find the exact value of t corresponding to Q.

c Hence find the

i coordinates of Q
ii value of k.

4.4 Tangents and normals

You can find the equation of a tangent or a normal to a parabola by using differentiation.

Refer to C1 and 1.5 for tangents and normals to a curve.

EXAMPLE 1

A parabola has equation $y^2 = 9x$
Find an equation for the tangent to the parabola at the point $P(4, -6)$.

The tangent has equation $y = mx + c$ where m is the gradient of the parabola at P.

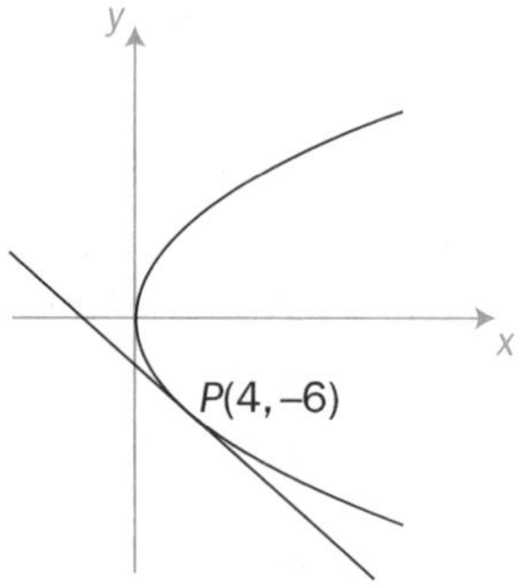

Make y the subject of the equation and then differentiate $y^2 = 9x$ with respect to x:

$$y^2 = 9x$$

$$y = \sqrt{9x}$$

$$= \pm 3x^{\frac{1}{2}}$$

so $$\frac{dy}{dx} = \pm\frac{3}{2}x^{-\frac{1}{2}}$$

Find m by substituting the x-coordinate of P into $\frac{dy}{dx}$:

When $x = 4$, $\frac{dy}{dx} = \pm\frac{3}{2} \times 4^{-\frac{1}{2}} = \pm\frac{3}{4}$

$4^{-\frac{1}{2}} = \frac{1}{4^{\frac{1}{2}}} = \frac{1}{2}$

At P the gradient of the curve is negative and so $m = -\frac{3}{4}$

Use the diagram to determine the correct sign for the gradient.

The equation of the tangent is $y = -\frac{3}{4}x + c$

Use the coordinates of P to find c:

You could also use $y - y_1 = m(x - x_1)$ – See C1

$$y = -\frac{3}{4}x + c$$

$$-6 = -\frac{3}{4} \times 4 + c$$

$$c = -3$$

Hence an equation for the tangent to this parabola at $P(4, -6)$ is $y = -\frac{3}{4}x - 3$

You can write the equation using integer coefficients as $4y + 3x + 12 = 0$

The gradient of the parabola with equation $y^2 = 4ax$ at any point $P(x, y)$, is given by $\frac{dy}{dx} = \pm\sqrt{\frac{a}{x}}$ where the sign of the gradient is the same as the sign of y.

e.g. When $y > 0$, P is on the upper half of the parabola and so the gradient is also positive.

EXAMPLE 2

A rectangular hyperbola has equation $xy = 18$
Find the gradient of the normal to this curve when $x = 6$.

The curve has equation $xy = 18$

Differentiate to find the gradient of the curve when $x = 6$:

$$xy = 18$$

$$y = \frac{18}{x} = 18x^{-1}$$

so $$\frac{dy}{dx} = -18x^{-2} = -\frac{18}{x^2}$$

When $x = 6$, $\frac{dy}{dx} = -\frac{18}{6^2} = -\frac{18}{36} = -\frac{1}{2}$

gradient of normal × gradient of tangent = −1 Refer to C1

Hence the gradient of the normal when $x = 6$ is 2.

The gradient of the rectangular hyperbola with equation $xy = c^2$ at any point $P(x, y)$ is given by $\frac{dy}{dx} = -\frac{c^2}{x^2}$

Since $xy = c^2$, $\frac{dy}{dx}$ is also given by $-\frac{y}{x}$

You can also describe the gradient of a parabola or rectangular hyperbola at any point using a parameter.

EXAMPLE 3

Express, in terms of t, the gradient of

a the parabola with equation $y^2 = 4ax$, where $a > 0$ at the point $P(at^2, 2at)$, $t > 0$

b the rectangular hyperbola with equation $xy = c^2$ at the point $Q\left(ct, \frac{c}{t}\right)$, $t \neq 0$

a The parabola has gradient $\pm\sqrt{\frac{a}{x}}$

If $t > 0$, $y > 0$. P lies on the upper half of the parabola, so the gradient is positive.

Gradient at $P = \sqrt{\frac{a}{x}}$

$$= \sqrt{\frac{a}{at^2}} = \frac{1}{t}$$

At P, $x = at^2$

b The rectangular hyperbola has gradient $-\frac{c^2}{x^2}$

Gradient at $Q = -\frac{c^2}{(ct)^2}$

$$= -\frac{c^2}{c^2t^2} = -\frac{1}{t^2}$$

At Q, $x = ct$
You could use

$$\frac{dy}{dx} = -\frac{y}{x} = -\frac{\left(\frac{c}{t}\right)}{ct} = -\frac{1}{t^2}$$

FP1

The gradient of the parabola with equation $y^2 = 4ax$, where $a > 0$, at the point $P(at^2, 2at)$, $t \neq 0$, is $\frac{1}{t}$

The gradient of the rectangular hyperbola with equation $xy = c^2$ at the point $Q\left(ct, \frac{c}{t}\right)$, $t \neq 0$, is $-\frac{1}{t^2}$

Exercise 4.4

1 Find an equation for the tangent and the normal to each curve at the given point P.

a The parabola with equation $y^2 = 4x$ at the point $P(4, 4)$.

b The rectangular hyperbola with equation $xy = 4$ at the point $P\left(\frac{1}{2}, 8\right)$.

c The parabola with equation $y^2 = 12x$ at the point $P\left(\frac{4}{3}, -4\right)$.

d The rectangular hyperbola with equation $xy = 6$ at the point $P\left(2\sqrt{3}, \sqrt{3}\right)$.

2 Points on a parabola C are given by $(2t^2, 4t)$ where t is any real number. Point P on C corresponds to the value $t = \frac{1}{2}$

a Show that an equation, T, for the tangent to C at point P is given by $y = 2x + 1$

b Find an equation for the normal, N, to C at point P. Give your answer in the form $ay + bx = c$ for integers a, b and c.

3 A rectangular hyperbola, W, has equation $xy = 12$

a Show that the gradient of the normal, N, to W at the point $P(2, 6)$, is $\frac{1}{3}$.

b Hence find an equation for N.

c Find the coordinates of the point Q where N intersects the curve W again.

4 A parabola, M, has equation $y^2 = kx$, where k is a positive constant. Point $P(2, -6)$ lies on M.

a Find the value of k.

b Show that an equation for the tangent, T, to M at point P is given by $2y + 3x + 6 = 0$

c Find the coordinates of the point Q, where T intersects the directrix of M.

5 In parametric form, the coordinates of any point on the rectangular hyperbola, C, are $\left(\sqrt{8}t, \frac{\sqrt{8}}{t}\right)$, $t \neq 0$.

P and Q are two points on C at which the gradient of C is $-\frac{1}{2}$. The x-coordinate of P is positive.

a Find the coordinates of the points P and Q.

b Find the coordinates of the two points on C at which the normals to C are parallel to the line PQ.

6 The diagram shows a parabola, E, with equation $y^2 = 24x$ and focus F. Point P on E has coordinates $\left(\frac{3}{2}, 6\right)$.

The line N is the normal to the curve E at point P. N intersects the x-axis at point Q.

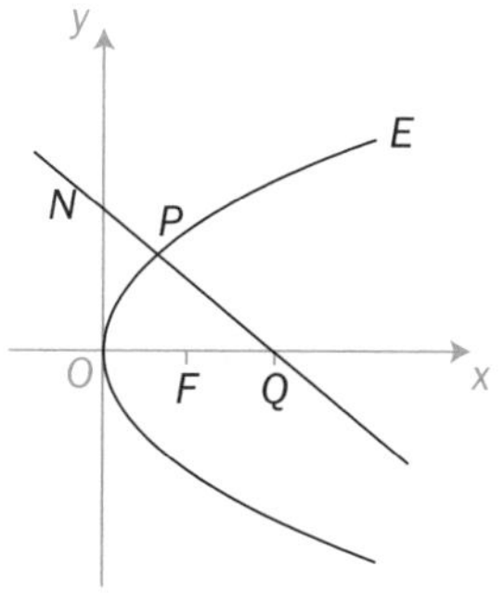

a Show that the equation of N can be expressed as $4y + 2x = 27$

b Hence find the coordinates of Q.

c Show that $\sin PFO = 0.8$ and hence find the area of triangle PFQ. You may use the result $\sin(180° - \alpha) \equiv \sin\alpha$

FP1

7 The diagram shows a rectangular hyperbola, C, with equation $xy = c^2$, for c a constant.

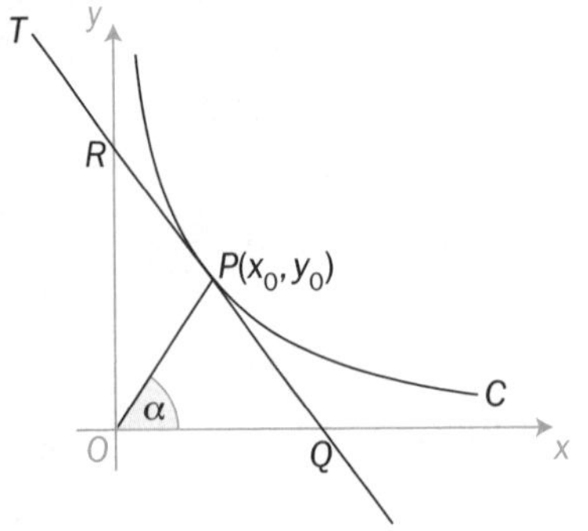

Point $P(x_0, y_0)$ is a point on C such that the line OP makes an angle α with the horizontal.
The tangent, T, to the curve at P crosses the x- and y-axes at points Q and R respectively.

a Show that an equation for T is given by $y = \frac{y_0}{x_0}(2x_0 - x)$

b Hence, or otherwise, find the coordinates of Q and R.

c Show that $x_0 \sin \alpha = y_0 \cos \alpha$

d Prove that triangles OPQ and OPR have equal areas.

8 The diagram shows the upper half of a parabola with equation $y^2 = 16x$ and directrix L.
The tangent, T, to this curve at point $P(x_0, y_0)$, intersects the directrix at point Q on the x-axis.

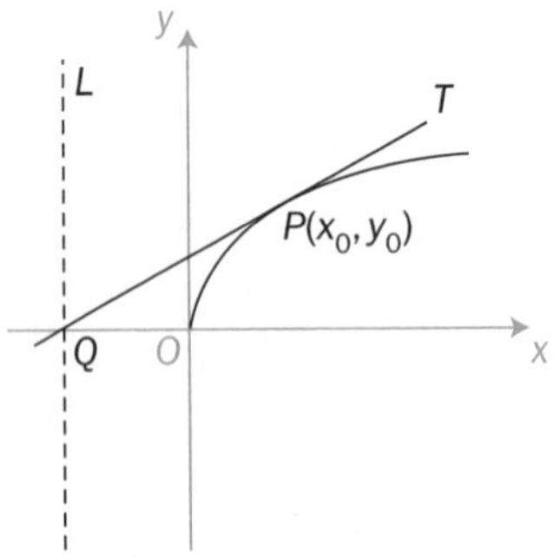

a Write down the coordinates of Q.

b Show that the gradient of T is $\frac{2}{\sqrt{x_0}}$

c Find the coordinates of point P.

9 A particular point, P, on the rectangular hyperbola H has coordinates $\left(ct, \frac{c}{t}\right)$, $t \neq 0, \pm 1$, where c is a positive constant.

a Show that the equation of the tangent to H at P is given by

$$t^2y = 2tc - x$$

b Hence write down the equation of the tangent to H at the point $Q\left(ct^2, \frac{c}{t^2}\right)$.

c Find, in terms of c and t, the coordinates of the point R, where these two tangents intersect.

10 The diagram shows a parabola, W.

The coordinates of any point, P, on W are $(at^2, 2at)$, where $a > 0$.

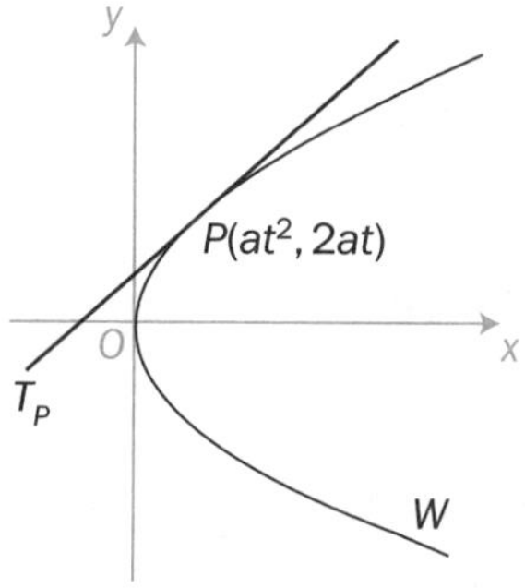

It is given that an equation for the tangent T_P to W at P is $ty = x + at^2$

a Write down, in terms of t, the gradient of T_P.

b Hence, or otherwise, show that an equation for the normal N to W at P is given by $y = at^3 + 2at - tx$

c i Find, in terms of a and t, the coordinates of the point Q on W at which the tangent T_Q to W is parallel to N.

ii Hence show that an equation for T_Q is $ty = -(t^2x + a)$

d Show that the lines T_P and T_Q intersect at a point on the directrix of W.

FP1

1 The focus, F, of a parabola, C, has coordinates $(2, 0)$.

a Find the equation of the parabola.

b Hence find the exact coordinates of the points P and Q where the line with equation $x = \frac{3}{2}$ intersects C.

c Show that the quadrilateral $OPFQ$ has area $4\sqrt{3}$.

2 A rectangular hyperbola, H, has equation $xy = 10$

a Find the coordinates of the points P and Q where the line with equation $y = 2x + 8$ intersects H. You may assume that P has a positive x-coordinate.

b Determine whether or not this line is the normal to H at either of the points P and Q.

3 In parametric form, the coordinates of any point on a rectangular hyperbola, C, are $\left(4t, \frac{4}{t}\right)$, $t \neq 0$.

a Show that an equation for the tangent, T, to C at the point P corresponding to $t = 2$ is given by $4y = 16 - x$

b Use algebra to prove that T does not intersect C again.

c Find the coordinates of the other point, Q, on C at which the tangent to C is parallel to T.

4 The diagram shows a parabola, M. In parametric form, the coordinates of any point on M are $(kt^2, 5t)$, for k a constant. Point F is the focus of M and point P on M corresponds to $t = \sqrt{2}$. The angle $OFP = \theta$

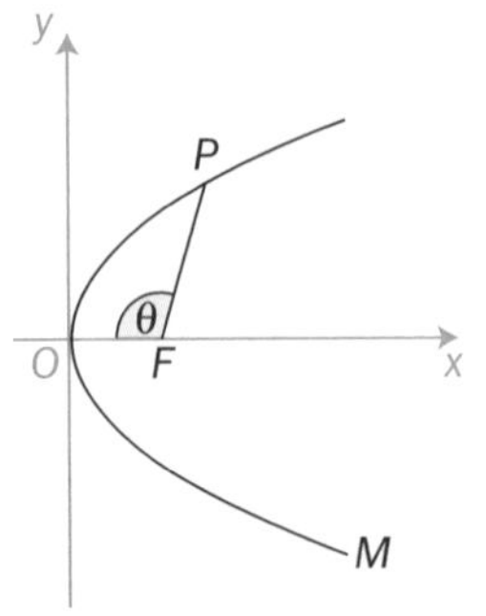

a Show that $k = 2.5$ and write down the coordinates of F.

b Find the x-coordinate of P and hence show that $FP = 7.5$

c Given that $OP = 5\sqrt{3}$ show that $\cos\theta = -\frac{1}{3}$

d Find the exact area of triangle OFP.

5 The diagram shows the parabola G with directrix L. Any point on G has coordinates $(t^2, 2t)$ for t a real number. Points P and Q on G correspond to the values $t = 3$ and $t = -3$ respectively.

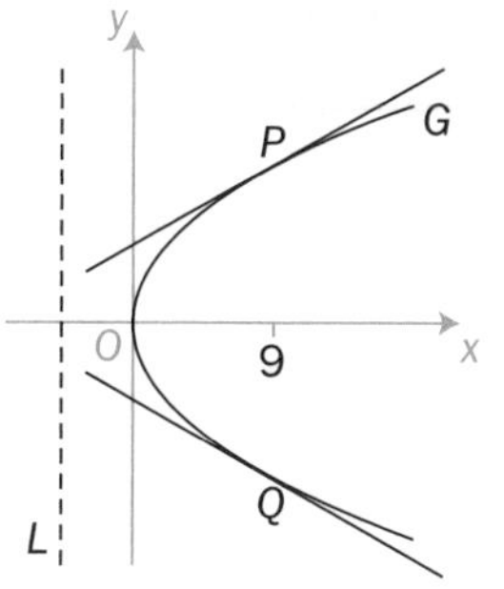

a Find equations for the tangents to the curve at each of the points P and Q.

b Show that these tangents intersect at the point $(k, 0)$, for k a constant to be found.

c Find the area of the region bounded by these tangents, the directrix and which does not contain the origin.

d Write down, in parametric form, the coordinates of any point on the parabola which has as its directrix the line $x = k$

6 The diagram shows a rectangular hyperbola with equation $xy = 12$. T_1 and T_2 are tangents to this curve at points P and Q respectively. The y-coordinate of P is 6 and the x-coordinate of Q is -18. T_1 and T_2 intersect at point R.

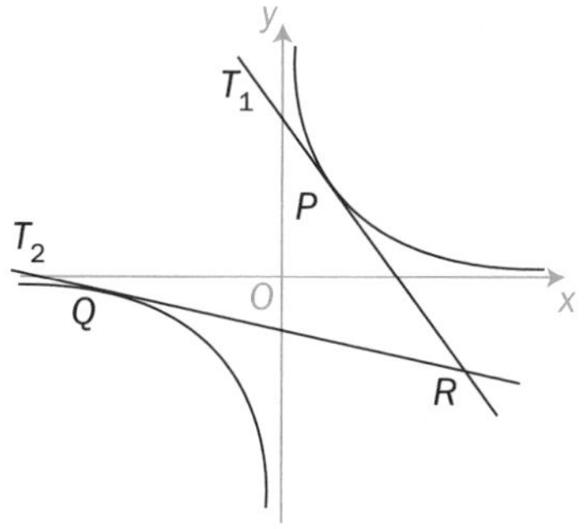

a Show that the equation of T_1 is $y = -3x + 12$ and find an equation for T_2.

b Hence find the coordinates of R.

c Calculate the gradient of the line PQ.

d Hence show that triangle PQR is right-angled and state the vertex at which the right-angle occurs.

FP1

7 A parabola, C, has equation $y^2 = 8x$ and focus F. Point P on C is such that the equation of the normal, N, to C has equation $y = \sqrt{2}(8 - x)$

a By considering the gradient of N, briefly explain why P must lie on the upper half of C and show that the coordinates of P are $(4, 4\sqrt{2})$.

b On the same diagram, sketch the parabola C and the line N.

c Find the exact coordinates of the other point Q where N intersects C again.

It is given that N crosses the x-axis at point R, where $PR = 4\sqrt{3}$

d i Find the coordinates of R.

ii Using the focus-directrix property of a parabola, or otherwise, show that triangle FPR is isosceles and hence find the exact value of $\cos P\hat{F}R$.

8 A rectangular hyperbola, H, has equation $xy = c^2$ where c is a positive constant. Use algebra to

a show that the line with equation $y = -x + 2c$ forms a tangent to H and find, in terms of c, the coordinates of its point of contact with H

b find, in terms of c, the range of values of k for which the line with equation $y = -x + k$ does not intersect H.

9 The diagram shows a parabola, V, with equation $y^2 = 4ax$ for $a > 0$ and with focus F. $P(x_0, y_0)$ is a point on the upper half of V such that $x_0 > a$. Line N is the normal to V at P and crosses the x-axis at point Q.

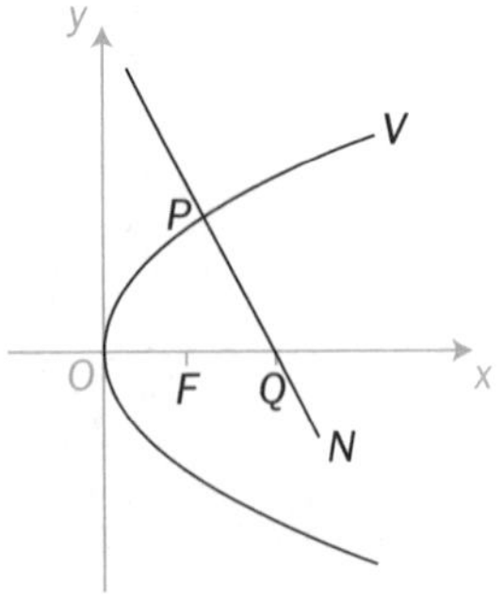

a Show that an equation for N is given by

$$y = 2\sqrt{ax_0} - \sqrt{\frac{x_0}{a}}(x - x_0)$$

b Hence find, in terms of x_0 and a, the coordinates of Q.

c Using the focus-directrix property, or otherwise, deduce that $FP = FQ$

d Given that $x_0 > a$, find, in terms of a, the value of x_0 for which triangle PFQ is equilateral.

10 The diagram shows part of a rectangular hyperbola with equation $xy = c^2$ for c a constant. P are Q are any two points on the curve, where $x_0, x_1 > 0$. Points R and S lie on the x-axis vertically below P and Q respectively.

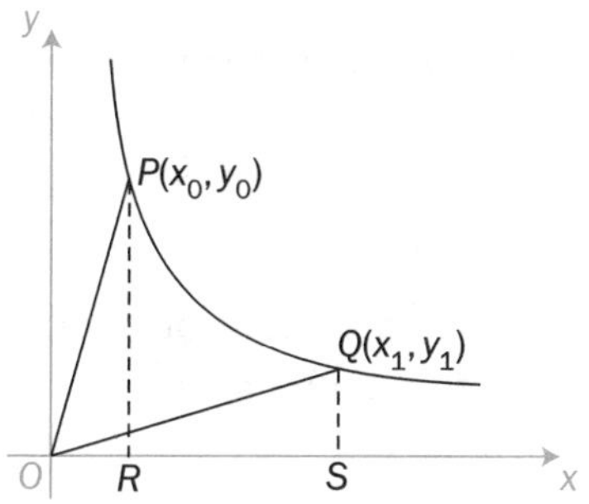

a Prove that triangles ORP and OSQ have equal areas.

b i Show that $(x_0 + y_0)^2 = OP^2 + 2c^2$

ii Deduce that if $OP = OQ$ then triangles ORP and OSQ have equal perimeters.

11 The point $P(at^2, 2at)$ for $t > 0$, lies on the parabola M with equation $y^2 = 4ax$, where a is a positive constant.

a Show that an equation for the tangent to M at P is $ty = x + at^2$.

The point $Q(16at^2, 8at)$ also lies on M.

b Use the result of **a** to write down an equation for the tangent to M at Q.

c Find, in terms of a and t, the coordinates of the point where these two tangents intersect.

12 **a** Show that the normal to the rectangular hyperbola $xy = c^2$, at the point $P\left(ct_0, \frac{c}{t_0}\right)$, $t \neq 0$ has equation

$$y = t_0^2 x + \frac{c}{t_0} - ct_0^3$$

The normal to the hyperbola at P meets the hyperbola again at the point Q.

b Show that the value of t corresponding to point Q satisfies the quadratic equation

$$t_0^3 t^2 + \left(1 - t_0^4\right)t - t_0 = 0$$

c Hence find, in terms of t_0, the coordinates of Q.

FP1

4 Exit

Summary

[Refer to]

- A parabola has cartesian equation $y^2 = 4ax$ where $a > 0$ is a constant. Its focus is at F, $(a, 0)$, and its directrix, L, has equation $x = -a$ 4.1
- Every point P on a parabola is equidistant from its focus and directrix. 4.1
- A rectangular hyperbola has cartesian equation $xy = c^2$ where $c > 0$ is a constant. 4.2
- Every point P on the parabola $y^2 = 4ax$ can be expressed in parametric form $P(at^2, 2at)$ where $t \in \mathbb{R}$ 4.3
- Every point P on the rectangular hyperbola $xy = c^2$ can be expressed in parametric form $P\left(ct, \frac{c}{t}\right)$ where t is any (non-zero) real number. 4.3
- The gradient of a parabola at any point $P(x, y)$ is $\pm\sqrt{\frac{a}{x}}$ where the sign of the gradient is given by the sign of y. In parametric form, the gradient of the parabola at point $P(at^2, 2at)$ is $\frac{1}{t}$ 4.4
- The gradient of a rectangular hyperbola at any point $P(x, y)$ is $-\frac{c^2}{x^2}$ In parametric form, the gradient of the hyperbola at point $P\left(ct, \frac{c}{t}\right)$, $t \neq 0$, is $-\frac{1}{t^2}$ 4.4

Links

Parabolic curves arise both in man-made constructions and in nature. They are the ideal shape when designing a satellite dish where radio waves need to be collected and then concentrated at a single point (the focus). The parabolic mirror in a headlight directs the particles of light from the source (which is at the focus) in one direction, allowing them to shine light at some distance straight ahead.

The path of a projectile, such as a golf ball, is parabolic in shape and can be simulated using appropriate software to predict maximum heights and ranges achieved for different launch velocities and angles.

The rectangular hyperbola occurs in many areas of research where two variables following an inverse relationship are being modelled.

Revision 1

1 The complex numbers z_1 and z_2 are given by $z_1 = 5 + 3i$ and $z_2 = 1 + pi$, where p is an integer.

a Find $\frac{z_2}{z_1}$ in the form $a + ib$, where a and b are expressed in terms of p.

b Given that $\arg\left(\frac{z_2}{z_1}\right) = \frac{1}{4}\pi$, find the value of p. [(c) Edexcel Limited 2007]

2 A rectangular hyperbola, R, has equation $xy = 6$

a Find the coordinates of the points where the line l with equation $y = 4x + 5$ intersects R.

b Show that an equation for the normal N to R at the point where $x = 3$ is $2y = 3x - 5$

c Find the coordinates of the point D where l and N meet and hence write down the cartesian equation of the rectangular hyperbola which passes through point D.

3 $f(x) = x^3 + x - 3$

The equation $f(x) = 0$ has a root, α, between $x = 1$ and $x = 2$

a By considering $f'(x)$, show that α is the only real root of the equation $f(x) = 0$

b Taking 1.2 as your first approximation to α, apply the Newton-Raphson procedure once to $f(x)$ to obtain a second approximation to α. Give your answer to three significant figures.

c Prove that your answer to part **b** gives the value of α correct to three significant figures. [(c) Edexcel Limited 2002]

4 A parabola, C, has equation $y^2 = 8x$ and focus F.

a Use algebra to find the coordinates of the points P and Q where the line with equation $y = 8 - 2x$ intersects C. You may assume P has a positive y-coordinate.

b Given that $\cos P\hat{F}Q = -\frac{4}{5}$ use the focus-directrix property of C to show that PQ has length $6\sqrt{5}$.

5 It is given that $z = 5 + 5i$ and $w = -2 + 6i$

a Find $|w|$ giving your answer in simplified surd form.

The complex numbers z and w are represented by points A and B on the Argand diagram.

b Show points A and B on an Argand diagram.

c Prove that triangle OAB is isosceles and show that $\cos A\hat{O}B = \frac{1}{5}\sqrt{5}$

d Show that $\frac{z}{w} = \frac{1}{2} - i$ and hence find $\arg\left(\frac{z}{w}\right)$, giving your answer in radians correct to two decimal places.

6 $f(x) = 4x^4 + 17x^2 + 4$

a Show that $2i$ is a root of the equation $f(x) = 0$

b Hence solve $f(x) = 0$ completely.

7 a By factorisation, show that two of the roots of the equation $x^3 - 27 = 0$ satisfy the equation $x^2 + 3x + 9 = 0$

b Hence, or otherwise, find the three cube roots of 27, giving your answers in the form $a + ib$, where $a, b \in \mathbb{R}$

c Show these roots on an Argand diagram. [(c) Edexcel Limited 2003]

8 $f(x) = 5x - 4\sqrt{x} - 2$

a Show that the equation $f(x) = 0$ has a root α between $x = 1.3$ and $x = 1.4$

b Starting with the interval [1.3, 1.4], use interval bisection three times to find an interval of width 0.0125 which contains α.

c Taking 1.3 as a first approximation to α, use the Newton-Raphson process on $f(x)$ once to obtain a second approximation to α. Give your answer to three decimal places.

9 The parabola C has equation $y^2 = 6x$. Point P on C has coordinates $P(6, 6)$

a Write down the coordinates of the focus, F, of C.

b Show that the equation of the tangent to C at point P is $y = \frac{1}{2}x + 3$

The tangent crosses the y-axis at point Q.

c Calculate the gradient of the line QF and deduce that triangle PQF is right-angled.

d Show that $QF = \frac{3}{2}\sqrt{5}$ and find the exact value of $\sin Q\hat{P}F$.

10 $z = -4 - 5i$

a Calculate arg z, giving your answer in radians to three decimal places.

The complex number w is given by $w = \frac{pi}{4 + 2i}$, where p is a positive constant.

b Show that $w = \frac{p}{20}(2 + 4i)$ and hence write down the exact value of arg w

c Given that $|w| = \frac{1}{2}\sqrt{5}$, find the value of p and hence express w in the form $a + ib$, where a and b are constants.

11 $f(x) = \frac{2x + 1}{x^2} - 2x, \quad x > 0$

The equation $f(x) = 0$ has exactly one real root, α, which lies in the interval $(1, 2)$.

a Using the end points of this interval find, by linear interpolation, an approximation to α. Give your answer to two decimal places.

b Show that $f'(x) = -2\left(\frac{1}{x^2} + \frac{1}{x^3} + 1\right)$

c Taking your answer to part **a** as a first approximation to α, apply the Newton-Raphson procedure once to $f(x)$ to find a second approximation to α.

d Show, by establishing a change of sign over an appropriate interval, that your answer to part **c** is accurate to two decimal places.

12 The diagram shows part of a rectangular hyperbola, V, with equation $xy = c^2$ for c a positive constant.

Points $P\left(ct, \frac{c}{t}\right)$ and $Q\left(2ct, \frac{c}{2t}\right)$, where $t > 0$, lie on C.

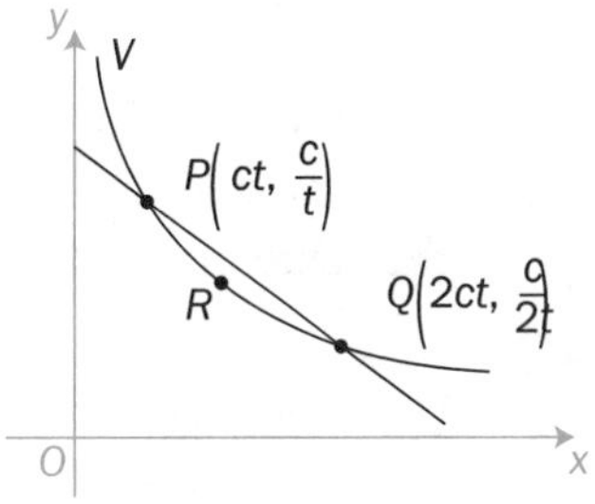

a Show that the gradient of the line PQ is given by $-\frac{1}{2t^2}$

Point R, shown in the diagram, lies on V.
The tangent to V at point R is parallel to the line PQ.

b Find, in terms of c and t, the coordinates of R.

13 Given that $3 + i$ is a root of the equation $f(x) = 0$, where $f(x) = 2x^3 + ax^2 + bx - 10, \quad a, b \in \mathbb{R}$

a find the other two roots of the equation $f(x) = 0$

b find the value of a and the value of b.

14 Given that $\frac{z + 2i}{z - \lambda i} = i$, where λ is a positive, real constant,

a show that $z = \left(\frac{1}{2}\lambda + 1\right) + i\left(\frac{1}{2}\lambda - 1\right)$

Given also that $\arg z = \arctan\frac{1}{2}$, calculate

b the value of λ

$\arctan\frac{1}{2} = \tan^{-1}\frac{1}{2}$

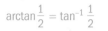

c the value of $|z|^2$.

[(c) Edexcel Limited 2006]

15 The diagram shows a rectangular hyperbola C.

The coordinates of any point on C are given by $\left(4t, \frac{4}{t}\right)$, $t \neq 0$

Line l with equation $5y = 8x + 24$ intersects C at points P and Q.

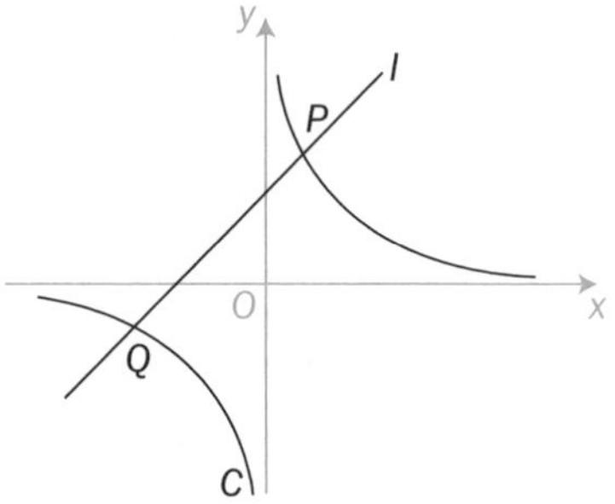

a Show that any value of t corresponding to a point $\left(4t, \frac{4}{t}\right)$ where l and C intersect satisfies the equation $8t^2 + 6t - 5 = 0$

b Solve this equation and hence find the coordinates of points P and Q.

Point $R\left(4t_0, \frac{4}{t_0}\right)$ on C, where $t_0 > \frac{1}{2}$, is such that angle $QPR = 90°$

c Show that t_0 satisfies the equation $10t_0^2 - 37t_0 + 16 = 0$ and hence find the coordinates of R.

16 $f(x) = \sqrt{x}(4 - x\sqrt{x}) - 2$

a Show that the equation $f(x) = 0$ has a root α in the interval $1.84 < x < 1.86$.

b Starting with this interval, use interval bisection twice to find an interval of width 0.005 which contains α and hence state the value of α to two decimal places.

You may assume α is the only root of the equation $f(x) = 0$ in the interval (1.84, 1.86).

c i Show that $f'(x) = 2x^{-\frac{1}{2}} - 2x$

ii Taking $x_1 = 2$ as a first approximation to α, apply the Newton-Raphson procedure once to $f(x)$ to obtain a second approximation to α.
State, with a reason, whether this value is an underestimate or overestimate for α.

d Explain, justifying your answer, why $x_1 = 1$ is not an appropriate starting value when applying the Newton-Raphson procedure to $f(x)$.

FP1

5 Matrix algebra

This chapter will show you how to
- apply the rules of matrix algebra
- find the determinant and, where possible, the inverse of a matrix
- represent linear transformations using matrices
- use the determinant as an area scale factor.

Before you start

You should know how to:

1 Work with transformations in the x–y plane.

e.g. Write down the coordinates of point Q which is the reflection of the point $P(4, 3)$, in the y-axis.

The y-axis is the mirror line:

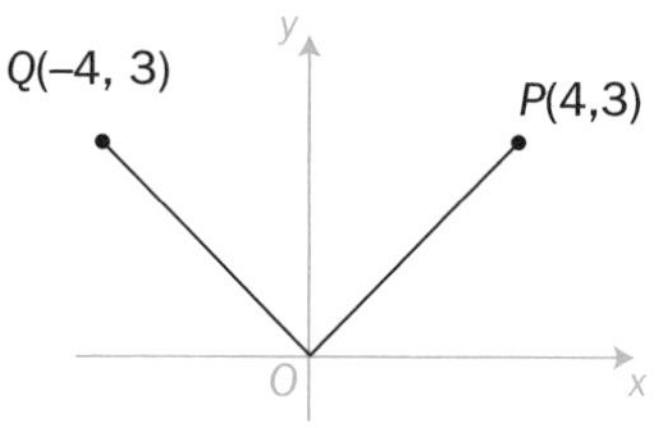

Q has coordinates $(-4, 3)$

2 Calculate the area of plane shapes.

e.g. Calculate the area of the parallelogram $ABCD$ where $AB = 10$, $AD = 6$ and angle $D\hat{A}B = 60°$

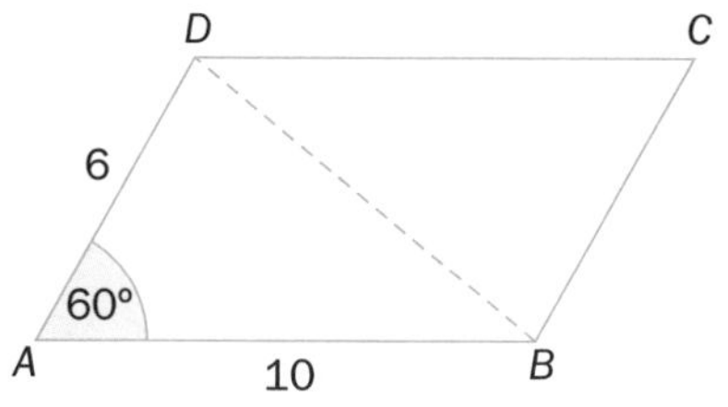

Area of $ABCD = 2 \times$ area of triangle ABD

$$= 2 \times \frac{1}{2} \times 10 \times 6 \times \sin 60°$$

$$= 60 \times \frac{\sqrt{3}}{2}$$

$$= 30\sqrt{3} \text{ square units}$$

Check in:

See GCSE for revision.

1 a Write down the coordinates of the point Q which is the result of applying each of the transformations to the point $P(4, 4)$.

i A reflection in the x-axis.

ii A reflection in the line $y = x$

iii A rotation of 180°, about the origin.

iv An enlargement, scale factor 1.25, centre the origin.

b Show that the transformation of an anticlockwise rotation of 90°, about the origin, followed by a reflection in the line $y = -x$ leaves the point $P(0, k)$ unchanged, where $k \in \mathbb{R}$.

2 a Find the area of Refer to C2.

i an equilateral triangle with sides of length 5 units

ii a parallelogram with adjacent sides 8 and 5 units enclosing an angle of 45°.

b Triangle OAB, where O is the origin, A is the point $(6, 0)$ and B is the point $(4, 2)$, is enlarged by scale factor 1.5, centre O.

i Find the coordinates of the vertices of the enlarged triangle.

ii Find the area of the enlarged triangle.

5.1 Definition of a matrix

A matrix is a collection of numbers arranged in rows and columns.

You can use a matrix to store efficiently information about a mathematical situation.

e.g You can describe the translation which maps the graph with equation $y = x^2$ onto the graph with equation $y = x^2 + 3$ either in words as

'move 0 units along the x-axis
and +3 units along the y-axis',

or, more efficiently, as the matrix $\begin{pmatrix} 0 \\ 3 \end{pmatrix}$

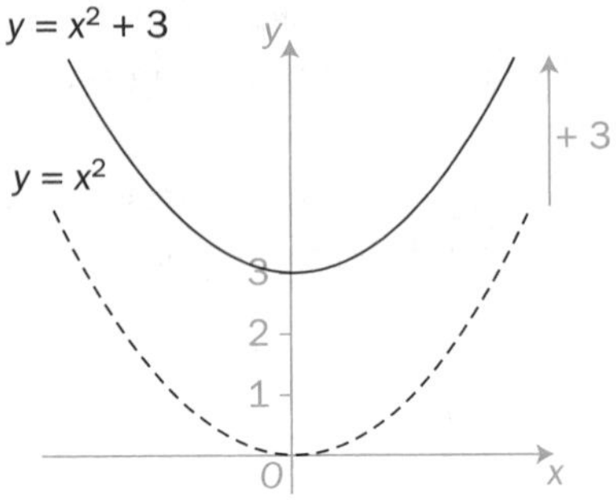

This matrix has two rows and one column

1 column ↓

1st row → 0
2nd row → 3

$\begin{pmatrix} 0 \\ 3 \end{pmatrix}$

so the matrix has order 2×1.

$\begin{pmatrix} 0 \\ 3 \end{pmatrix}$ is also a vector.

A matrix with r rows and c columns has order $r \times c$

The quantities inside a matrix are called its **elements**.
You can refer to the elements of a matrix using its rows and columns.

e.g In the 2×2 matrix, $\begin{pmatrix} 4 & -5 \\ \frac{1}{2} & \pi \end{pmatrix}$, the element in

row 1, column 1 is 4
row 1, column 2 is -5
row 2, column 1 is $\frac{1}{2}$
row 2, column 2 is π

Column 1 ↓ Column 2 ↓
Row 1 → 4 −5
Row 2 → $\frac{1}{2}$ π

You can use a bold capital letter to represent a matrix.
The elements of the matrix are then referred to using the corresponding lower-case letter.

e.g. If $\mathbf{A} = \begin{pmatrix} 3 & -5 \\ \frac{1}{2} & \pi \end{pmatrix}$ then the element in row 1 column 2 is denoted by a_{12}.

Hence $a_{12} = -5$ and $a_{21} = \frac{1}{2}$ etc.

The element in the ith row and jth column of matrix A is a_{ij}
The notation $\mathbf{A}_{r \times c}$ means matrix A has order $r \times c$

EXAMPLE 1

a Write down the order of B.
b Write down the value of b_{31}.
c Write down two elements which are equal.

$$B = \begin{pmatrix} 7 & 4 \\ 3 & 0 \\ 5 & 7 \end{pmatrix}$$

---

a B has 3 rows and 2 columns Hence B has order 3 × 2.

You could also write $B_{3 \times 2}$.

b b_{31} is the element in the 3rd row and 1st column.

$$\begin{pmatrix} 7 & 4 \\ 3 & 0 \\ 5 & 7 \end{pmatrix} \qquad \text{Hence } b_{31} = 5$$

c The two elements which are equal are b_{11} and b_{32}.

Each of these elements has value 7.

Exercise 5.1

1 Write down the order of each matrix.

a $\begin{pmatrix} 4 & 1 & 0 \\ 2 & 5 & 7 \end{pmatrix}$ **b** $\begin{pmatrix} 5 \\ 3 \\ 1 \end{pmatrix}$ **c** $\begin{pmatrix} 1 & 1 & 2 & 3 \end{pmatrix}$ **d** $\begin{pmatrix} -1 & 3 & 6 \\ 0 & 4 & 2 \\ 2 & 1 & 5 \end{pmatrix}$

2 Given that $A = \begin{pmatrix} 5 & \frac{1}{3} & 0 \\ \frac{2}{3} & -5 & 3 \end{pmatrix}$, write down the value of

a a_{23} **b** $a_{12} + a_{21}$ **c** $a_{11}a_{22}$

3 The elements of a particular 3 × 3 matrix, **B**, satisfy the equation $b_{ij} = b_{ji}$ for all values $1 \leqslant i \leqslant 3$, $1 \leqslant j \leqslant 3$.

Find all the elements of **B** given that $B = \begin{pmatrix} 2 & 1 & \square \\ \square & 7 & \frac{3}{4} \\ 6 & \square & 0 \end{pmatrix}$

4 The elements of a particular 4 × 4 matrix, **C**, satisfy the equation $c_{ij} + c_{ji} = 0$ for all values $1 \leqslant i \leqslant 4$, $1 \leqslant j \leqslant 4$.

Find all the elements of **C** given that $C = \begin{pmatrix} \square & \square & \pi & \square \\ -1 & \square & 2 & 0 \\ \square & \square & \square & \square \\ \sqrt{2} & \square & 7 & \square \end{pmatrix}$

5 The elements of a particular 2 × 2 matrix, **A**, satisfy the equation $a_{ij} = 2a_{ji}^2$ for all values $1 \leqslant i \leqslant 2$, $1 \leqslant j \leqslant 2$.

Given that all of its elements are non-zero, find the matrix **A**.

FP1

5.2 Matrix algebra

Adding and subtracting matrices

You can add or subtract two matrices, if they have the same order, by adding or subtracting corresponding elements.

Refer to Section 5.1 for the definition of order.

You can multiply any matrix by a constant k by multiplying each of its elements by k.

EXAMPLE 1

If $\mathbf{A} = \begin{pmatrix} 1 & 4 & 2 \\ -1 & 3 & 2 \end{pmatrix}$ and $\mathbf{B} = \begin{pmatrix} 2 & 4 & 0 \\ 5 & -2 & -1 \end{pmatrix}$ find

a $\mathbf{A}+\mathbf{B}$ **b** $3\mathbf{A}$ **c** $\mathbf{A}-2\mathbf{B}$

You can add **A** and **B** since they have the same order.

a $\mathbf{A}+\mathbf{B} = \begin{pmatrix} 1 & 4 & 2 \\ -1 & 3 & 2 \end{pmatrix} + \begin{pmatrix} 2 & 4 & 0 \\ 5 & -2 & -1 \end{pmatrix}$

$= \begin{pmatrix} 1+2 & 4+4 & 2+0 \\ -1+5 & 3+(-2) & 2+(-1) \end{pmatrix}$

$= \begin{pmatrix} 3 & 8 & 2 \\ 4 & 1 & 1 \end{pmatrix}$

Add corresponding elements.

The answer has the same order as each of **A** and **B**.

b $3\mathbf{A} = 3\times\begin{pmatrix} 1 & 4 & 2 \\ -1 & 3 & 2 \end{pmatrix} = \begin{pmatrix} 3\times1 & 3\times4 & 3\times2 \\ 3\times(-1) & 3\times3 & 3\times2 \end{pmatrix}$

$= \begin{pmatrix} 3 & 12 & 6 \\ -3 & 9 & 6 \end{pmatrix}$

Multiply each element in **A** by 3.

c $\mathbf{A}-2\mathbf{B} = \begin{pmatrix} 1 & 4 & 2 \\ -1 & 3 & 2 \end{pmatrix} - 2\begin{pmatrix} 2 & 4 & 0 \\ 5 & -2 & -1 \end{pmatrix}$

$= \begin{pmatrix} 1 & 4 & 2 \\ -1 & 3 & 2 \end{pmatrix} - \begin{pmatrix} 4 & 8 & 0 \\ 10 & -4 & -2 \end{pmatrix}$

$= \begin{pmatrix} -3 & -4 & 2 \\ -11 & 7 & 4 \end{pmatrix}$

Work out the elements of 2**B** first.

Subtract each element of 2**B** from the corresponding element of **A**.

The zero matrix **O** has all its elements equal to zero.

e.g. $\mathbf{O}_{2\times 2} = \begin{pmatrix} 0 & 0 \\ 0 & 0 \end{pmatrix}$

$\mathbf{A} + \mathbf{O} = \mathbf{A} = \mathbf{O} + \mathbf{A}$ for any 2×2 matrix **A**

e.g. $\begin{pmatrix} 2 & 1 \\ 3 & 1 \end{pmatrix} + \begin{pmatrix} 0 & 0 \\ 0 & 0 \end{pmatrix} = \begin{pmatrix} 2 & 1 \\ 3 & 1 \end{pmatrix} = \begin{pmatrix} 0 & 0 \\ 0 & 0 \end{pmatrix} + \begin{pmatrix} 2 & 1 \\ 3 & 1 \end{pmatrix}$

The zero matrix **O** behaves like the number 0 in ordinary addition.

Two matrices **A** and **B** are equal if their orders are equal and $a_{ij} = b_{ij}$ for all possible elements.

EXAMPLE 2

Matrices $\mathbf{A} = \begin{pmatrix} 1 & p \\ q & 2v \end{pmatrix}$ and $\mathbf{B} = \begin{pmatrix} u & -3 \\ \frac{3}{2} & v \end{pmatrix}$,

where p, q, u and v are real numbers, are such that **A** = **B**.

a Find the values of p, q, u and v.

b Find the matrix **C** such that **A** + **B** + **C** = **O**

---

a If $\begin{pmatrix} 1 & p \\ q & 2v \end{pmatrix} = \begin{pmatrix} u & -3 \\ \frac{3}{2} & v \end{pmatrix}$ then corresponding elements are equal.

Compare corresponding elements:

$u = 1, p = -3, q = \frac{3}{2}$ and $2v = v \Rightarrow v = 0$

b $\mathbf{A} = \begin{pmatrix} 1 & -3 \\ \frac{3}{2} & 0 \end{pmatrix} = \mathbf{B}$

So $\mathbf{A} + \mathbf{B} = \begin{pmatrix} 1 & -3 \\ \frac{3}{2} & 0 \end{pmatrix} + \begin{pmatrix} 1 & -3 \\ \frac{3}{2} & 0 \end{pmatrix} = \begin{pmatrix} 2 & -6 \\ 3 & 0 \end{pmatrix}$

If **A** + **B** + **C** = **O**

then $\begin{pmatrix} 2 & -6 \\ 3 & 0 \end{pmatrix} + \mathbf{C} = \begin{pmatrix} 0 & 0 \\ 0 & 0 \end{pmatrix}$

So $\mathbf{C} = \begin{pmatrix} 0 & 0 \\ 0 & 0 \end{pmatrix} - \begin{pmatrix} 2 & -6 \\ 3 & 0 \end{pmatrix} = \begin{pmatrix} -2 & 6 \\ -3 & 0 \end{pmatrix}$

Subtract = $\begin{pmatrix} 2 & -6 \\ 3 & 0 \end{pmatrix}$ from both sides of the equation, as in normal algebra.

Hence $\mathbf{C} = \begin{pmatrix} -2 & 6 \\ -3 & 0 \end{pmatrix}$

FP1

Multiplying matrices

> You can only multiply two matrices if they are compatible, that is, if the number of columns in the left-hand matrix is the same as the number of rows in the right-hand matrix.

If matrix **A** has order $c \times d$ and matrix **B** has order $d \times e$ they are compatible and the product **AB** has order $c \times e$

A × B

$c \times d$ $d \times e$

same so multiplication is possible

AB has order $c \times e$

AB means **A** × **B**, as in normal algebra.

EXAMPLE 3

If $\mathbf{A} = \begin{pmatrix} 2 & -1 \\ 4 & 3 \\ 0 & 2 \end{pmatrix}$ and $\mathbf{B} = \begin{pmatrix} -2 & 1 \\ 2 & -1 \end{pmatrix}$ find the product **AB**.

Matrix **A** has order 3×2 and matrix **B** has order 2×2 so they are compatible. The product **AB** will have order 3×2.

A has 2 columns.
B has 2 rows.

$$\mathbf{A} \times \mathbf{B} = \begin{pmatrix} 2 & -1 \\ 4 & 3 \\ 0 & 2 \end{pmatrix} \times \begin{pmatrix} -2 & 1 \\ 2 & -1 \end{pmatrix}$$

3×2 2×2

Matrices are compatible

$$= \begin{pmatrix} (2 \times -2) + (-1 \times 2) & (2 \times 1) + (-1 \times -1) \\ (4 \times -2) + (3 \times 2) & (4 \times 1) + (3 \times -1) \\ (0 \times -2) + (2 \times 2) & (0 \times 1) + (2 \times -1) \end{pmatrix}$$

$$= \begin{pmatrix} -6 & 3 \\ -2 & 1 \\ 4 & -2 \end{pmatrix} \quad \text{a } 3 \times 2 \text{ matrix.}$$

$$\text{Hence } \mathbf{AB} = \begin{pmatrix} -6 & 3 \\ -2 & 1 \\ 4 & -2 \end{pmatrix}$$

Multiply the first element in row 1 of matrix **A** by the first element in column 1 of matrix **B**. (2×-2)

Multiply the second element in row 1 of matrix **A** by the second element in column 1 of matrix **B**. (-1×2)

Add these two answers to give the element in the first row and first column of matrix AB. $(2 \times -2) + (-1 \times 2) = -6$.

Continue in this way until each row in **A** has multiplied the corresponding column in **B**.

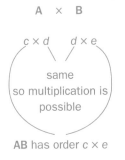

EXAMPLE 4

Given that $\mathbf{A} = \begin{pmatrix} 1 & -3 \\ 5 & 6 \\ -1 & 2 \end{pmatrix}$ and $\mathbf{B} = (2 \quad 0 \quad 4)$

state which of the products **AB** and **BA** is defined.
Find this product.

A has order 3 × 2, **B** has order 1 × 3.

Hence only the product **BA** is defined and has order 1 × 2.

B has 3 columns, **A** has 3 rows.

$$\mathbf{BA} = (2 \quad 0 \quad 4)\begin{pmatrix} 1 & -3 \\ 5 & 6 \\ -1 & 2 \end{pmatrix}$$

$$= ((2 \times 1) + (0 \times 5) + (4 \times -1) \quad (2 \times -3) + (0 \times 6) + (4 \times 2))$$
$$= (-2 \quad 2)$$

(−2 2) is a 1 × 2 matrix.

A × O = O = O × A for any 2 × 2 matrix **A**

e.g. $\begin{pmatrix} 2 & 1 \\ 3 & 1 \end{pmatrix} \times \begin{pmatrix} 0 & 0 \\ 0 & 0 \end{pmatrix} = \begin{pmatrix} 0 & 0 \\ 0 & 0 \end{pmatrix} = \begin{pmatrix} 0 & 0 \\ 0 & 0 \end{pmatrix} \times \begin{pmatrix} 2 & 1 \\ 3 & 1 \end{pmatrix}$

The zero matrix **O** behaves like the number 0 in ordinary multiplication.

FP1

A matrix is square if it has an equal number of rows and columns.

e.g. Any 2 × 2 matrix is a square matrix.

If **A** is a square matrix then the notation $\mathbf{A}^n$, where n is a positive integer, means **A × A × A … × A** n times.

EXAMPLE 5

If $\mathbf{A} = \begin{pmatrix} 1 & 2 \\ 3 & 4 \end{pmatrix}$ and $\mathbf{B} = \begin{pmatrix} 5 \\ 6 \end{pmatrix}$ find the product

a **AB** **b** $\mathbf{A}^2$

a **A** has order 2 × 2, **B** has order 2 × 1

Hence the product **AB** has order 2 × 1

2 × 2 2 × 1 — same

Work out each entry by summing the products of appropriate row elements of **A** with column elements of **B**:

$$\mathbf{AB} = \begin{pmatrix} 1 & 2 \\ 3 & 4 \end{pmatrix}\begin{pmatrix} 5 \\ 6 \end{pmatrix} = \begin{pmatrix} (1\times 5) + (2\times 6) \\ (3\times 5) + (4\times 6) \end{pmatrix} = \begin{pmatrix} 17 \\ 39 \end{pmatrix}$$

In this question the product **BA** cannot be found since **B** has 1 column and **A** has 2 rows.

b $$\mathbf{A}^2 = \begin{pmatrix} 1 & 2 \\ 3 & 4 \end{pmatrix}\begin{pmatrix} 1 & 2 \\ 3 & 4 \end{pmatrix} = \begin{pmatrix} (1\times 1) + (2\times 3) & (1\times 2) + (2\times 4) \\ (3\times 1) + (4\times 3) & (3\times 2) + (4\times 4) \end{pmatrix}$$

$$= \begin{pmatrix} 7 & 10 \\ 15 & 22 \end{pmatrix}$$

$\mathbf{A}^2$ means **A** multiplied with itself, as in normal algebra.

For any matrices A, B, C of compatible orders,

$\mathbf{A}(\mathbf{BC}) = (\mathbf{AB})\mathbf{C}$

This means you can find the product **ABC** by calculating either **A** (**BC**) or (**AB**) **C**.

The identity matrix, **I**, with entries a_{ij} is such that

$a_{ij} = 1$ if $i = j$ and $a_{ij} = 0$ if $i \neq j$

e.g. $\mathbf{I}_{22} = \begin{pmatrix} 1 & 0 \\ 0 & 1 \end{pmatrix}$

The elements a_{ij} such that $i = j$ form the leading diagonal of a matrix **A**.

Exercise 5.2

1 Find

a $\begin{pmatrix} 2 & 3 & 1 \\ 4 & -4 & 0 \\ 6 & 2 & 5 \end{pmatrix} + \begin{pmatrix} 3 & -1 & 2 \\ 8 & 2 & -5 \\ 9 & 7 & 6 \end{pmatrix}$ **b** $\begin{pmatrix} 3 & -5 \\ \frac{1}{3} & -7 \\ 2 & 1 \end{pmatrix} - \begin{pmatrix} 1 & 5 \\ \frac{2}{3} & 3 \\ 8 & -1 \end{pmatrix}$ **c** $\begin{pmatrix} 2 & 6 \\ 4 & 2 \end{pmatrix} + \begin{pmatrix} 3 & -1 \\ 1 & -7 \end{pmatrix} - \begin{pmatrix} 5 & 5 \\ 5 & 5 \end{pmatrix}$

FP1

2 $\mathbf{A} = \begin{pmatrix} 2 & 4 \\ 3 & 1 \end{pmatrix}$, $\mathbf{B} = \begin{pmatrix} \frac{1}{3} & \frac{4}{3} \\ 1 & 0 \end{pmatrix}$ and $\mathbf{C} = \begin{pmatrix} 4 & 5 \\ 6 & -6 \\ -2 & 4 \end{pmatrix}$

a Find **i** $\mathbf{A} + 3\mathbf{B}$ **ii** $-\frac{1}{2}\mathbf{C}$

iii the value of the constant k such that $\mathbf{A} + k\mathbf{B} = \mathbf{I}$, where **I** is the identity matrix.

b Explain why is not possible to find $\mathbf{A} + \mathbf{B} + \mathbf{C}$.

3 Matrices $\mathbf{A} = \begin{pmatrix} 8 & x^3 \\ y^2 & 3 \end{pmatrix}$ and $\mathbf{B} = \begin{pmatrix} 2p & -8 \\ 2y & \sqrt{q} \end{pmatrix}$

where p, q, x and y are constants, are such that $\mathbf{A} = \mathbf{B}$

a Find the values of p, q and x.

b Find the two possible values of y.

4 Find the following products.

a $\begin{pmatrix} 2 & 3 \\ 4 & 2 \end{pmatrix}\begin{pmatrix} 3 \\ 4 \end{pmatrix}$ **b** $\begin{pmatrix} -3 & 6 \\ 2 & \frac{1}{2} \end{pmatrix}\begin{pmatrix} 5 \\ -2 \end{pmatrix}$ **c** $(4 \quad -2)\begin{pmatrix} 1 & 5 & -3 \\ 3 & 7 & -2 \end{pmatrix}$

d $\begin{pmatrix} 4 & 2 \\ 7 & 5 \end{pmatrix}\begin{pmatrix} 5 & 3 \\ -2 & 6 \end{pmatrix}$ **e** $\begin{pmatrix} 3 & 5 \\ 0 & \frac{1}{2} \\ -\frac{1}{2} & 8 \end{pmatrix}\begin{pmatrix} 4 \\ 6 \end{pmatrix}$ **f** $\begin{pmatrix} 1 & -2 & 3 \\ 2 & -3 & 1 \\ 3 & 1 & 2 \end{pmatrix}\begin{pmatrix} 4 \\ 2 \\ 1 \end{pmatrix}$

5 **a** Given that $\mathbf{P} = \begin{pmatrix} 5 & 1 \\ 2 & 0 \end{pmatrix}$, $\mathbf{Q} = \begin{pmatrix} -2 & 0 \\ 3 & 4 \end{pmatrix}$ and $\mathbf{R} = \begin{pmatrix} 1 & -3 \\ 0 & 2 \end{pmatrix}$ find

i **PQ** ii **RP**

b Using part **a** find **PQRP**.

6 Matrices $\mathbf{A}_{2\times3}$ and $\mathbf{B}_{r\times c}$ are such that the product **AB** has order 2×1

a State the order of **B**.

b Determine whether or not the product **BA** is defined.

7 It is given that $\mathbf{A} = \begin{pmatrix} 2 & -1 \\ 1 & -2 \end{pmatrix}$

a Show that $\mathbf{A}^2 = 3\mathbf{I}$, where **I** is the 2×2 identity matrix.

b Express $\mathbf{A}^3$ in terms of **A**.

c Hence write down $\mathbf{A}^5$ in terms of **A**.

8 **a** Given that $\begin{pmatrix} 3 & 2 \\ 7 & -3 \end{pmatrix}\begin{pmatrix} p \\ q \end{pmatrix} = \begin{pmatrix} 5 \\ 27 \end{pmatrix}$, for constants p and q,

i form a pair of simultaneous equations in p and q
ii solve these equations to find the value of p and the value of q.

b Use a similar method to find the values of x and y

such that $\begin{pmatrix} 7 & -5 \\ 5 & -3 \end{pmatrix}\begin{pmatrix} x \\ y \end{pmatrix} = \begin{pmatrix} 21 \\ 17 \end{pmatrix}$

9 Given that $\mathbf{A} = \begin{pmatrix} 2 & x \\ 3 & y \end{pmatrix}$, find the values of the constants x and y such that $\mathbf{A}^2 = \mathbf{O}$

O is the 2×2 zero matrix.

10 For $\mathbf{C} = \begin{pmatrix} 1 & 0 \\ 3 & 2 \end{pmatrix}$

a verify that **CD** = **DC** where $\mathbf{D} = \begin{pmatrix} 4 & 0 \\ -3 & 3 \end{pmatrix}$

b show that if **CE** = **EC**, where $\mathbf{E} = \begin{pmatrix} a & b \\ c & d \end{pmatrix}$, then

a, b, c and d are integers.

i $b = 0$
ii c is a multiple of 3.

FP1

5.3 The determinant of a 2 × 2 matrix

The determinant of the matrix $\mathbf{A} = \begin{pmatrix} a & b \\ c & d \end{pmatrix}$ is $ad - bc$

The determinant of matrix **A** is written as det **A**, |**A**|, or as $\left|\begin{pmatrix} a & b \\ c & d \end{pmatrix}\right|$

EXAMPLE 1

Calculate the determinant of the matrix $\mathbf{A} = \begin{pmatrix} 4 & 1 \\ 2 & 3 \end{pmatrix}$

Compare $\begin{pmatrix} 4 & 1 \\ 2 & 3 \end{pmatrix}$ with $\begin{pmatrix} a & b \\ c & d \end{pmatrix}$

$$\det \mathbf{A} = ad - bc$$
$$= 4 \times 3 - 1 \times 2$$
$$= 10$$

Hence $\det \mathbf{A} = 10$

In Section 5.5 you will learn the geometrical significance of a determinant.

EXAMPLE 2

Find the value of x such that $\left|\begin{pmatrix} 7 & -3 \\ 0.5 & x \end{pmatrix}\right| = -23$

A determinant can be negative.

Compare $\begin{pmatrix} 7 & -3 \\ 0.5 & x \end{pmatrix}$ with $\begin{pmatrix} a & b \\ c & d \end{pmatrix}$

$$\left|\begin{pmatrix} 7 & -3 \\ 0.5 & x \end{pmatrix}\right| = ad - bc$$
$$= 7x - (-3) \times 0.5$$
$$= 7x + 1.5$$

Hence $7x + 1.5 = -23$

$$\therefore x = -3.5$$

If the determinant of a matrix is zero, the matrix is singular.
A non-singular matrix is one whose determinant is non-zero.

EXAMPLE 3

Given that $\mathbf{A} = \begin{pmatrix} 2 & 1 \\ 3 & 2 \end{pmatrix}$ and $\mathbf{B} = \begin{pmatrix} 3 & 2 \\ 6 & 4 \end{pmatrix}$

classify each of the following matrices as singular or non-singular.

a $\mathbf{A}$ **b** $\mathbf{B}$ **c** $\mathbf{B} - 2\mathbf{A}$

---

a $\mathbf{A} = \begin{pmatrix} 2 & 1 \\ 3 & 2 \end{pmatrix}$

$\det \mathbf{A} = 2 \times 2 - 1 \times 3 = 1$

Hence $\mathbf{A}$ is non-singular.

b $\mathbf{B} = \begin{pmatrix} 3 & 2 \\ 6 & 4 \end{pmatrix}$

$\det \mathbf{B} = 3 \times 4 - 2 \times 6 = 12 - 12 = 0$

Hence $\mathbf{B}$ is singular.

c $\mathbf{B} - 2\mathbf{A} = \begin{pmatrix} 3 & 2 \\ 6 & 4 \end{pmatrix} - 2\begin{pmatrix} 2 & 1 \\ 3 & 2 \end{pmatrix}$

$= \begin{pmatrix} 3-4 & 2-2 \\ 6-6 & 4-4 \end{pmatrix}$

$= \begin{pmatrix} -1 & 0 \\ 0 & 0 \end{pmatrix}$

$\det(\mathbf{B} - 2\mathbf{A}) = -1 \times 0 - 0 \times 0$

$= 0$

and so $\mathbf{B} - 2\mathbf{A}$ is a singular matrix.

Work out the elements of **B** – 2**A** first.

If $\mathbf{A}$ and $\mathbf{B}$ are any pair of 2×2 matrices then

$\det(\mathbf{AB}) = \det \mathbf{A} \times \det \mathbf{B}$

The product **AB** is defined since **A** and **B** have equal orders – refer to Section 5.2.

EXAMPLE 4

Given that $\mathbf{A} = \begin{pmatrix} 1 & 4 \\ -1 & 2 \end{pmatrix}$ and $\mathbf{AB} = \begin{pmatrix} 7 & 19 \\ 5 & 5 \end{pmatrix}$ find $\det \mathbf{B}$.

$\mathbf{A} = \begin{pmatrix} 1 & 4 \\ -1 & 2 \end{pmatrix}$ so $\det \mathbf{A} = 1 \times 2 - 4 \times (-1) = 6$

$\mathbf{AB} = \begin{pmatrix} 7 & 19 \\ 5 & 5 \end{pmatrix}$ so $\det(\mathbf{AB}) = 7 \times 5 - 19 \times 5 = -60$

$\det(\mathbf{AB}) = \det \mathbf{A} \times \det \mathbf{B}$

$-60 = 6 \times \det \mathbf{B}$

Hence $\det \mathbf{B} = -10$

FP1

Exercise 5.3

1 Calculate these determinants.

a $\begin{vmatrix} 2 & 5 \\ 1 & 4 \end{vmatrix}$ **b** $\begin{vmatrix} 4 & 1 \\ -2 & 3 \end{vmatrix}$

c $\begin{vmatrix} \frac{1}{2} & \frac{1}{3} \\ 6 & 8 \end{vmatrix}$ **d** $\begin{vmatrix} 3 & -2 \\ 5 & -4 \end{vmatrix}$

2 Given that $\mathbf{A} = \begin{pmatrix} 1 & 2 \\ -3 & 2 \end{pmatrix}$ and $\mathbf{B} = \begin{pmatrix} 3 & 1 \\ -2 & 1 \end{pmatrix}$

a Find **i** $\mathbf{A} + \mathbf{B}$ **ii** $\det(\mathbf{A} + \mathbf{B})$

b Show that $\det(2\mathbf{A} - 3\mathbf{B}) = -7$

c Verify that $\det(\mathbf{A} + \mathbf{B}) \neq \det \mathbf{A} + \det \mathbf{B}$

3 Solve these equations.

a $\begin{vmatrix} 2 & 1 \\ 5 & x \end{vmatrix} = 9$ **b** $\begin{vmatrix} 4 & x \\ -2 & x \end{vmatrix} = 3$

c $\begin{vmatrix} x & 2 \\ 3 & x \end{vmatrix} = 10$ **d** $\begin{vmatrix} 2x & 2 \\ x & x \end{vmatrix} = 12$

e $\begin{vmatrix} 2x & 4 \\ 1 & 3x \end{vmatrix} + 5x = 0$ **f** $\begin{vmatrix} x+1 & -3 \\ 2 & x-1 \end{vmatrix} = 6x$

4 Classify each matrix as singular or non-singular.

a $\begin{pmatrix} 8 & 3 \\ 5 & 2 \end{pmatrix}$ **b** $\begin{pmatrix} -6 & 12 \\ \frac{9}{2} & -9 \end{pmatrix}$

c $\begin{pmatrix} 21 & 18 \\ 14 & 12 \end{pmatrix}$ **d** $\begin{pmatrix} 0 & 5 \\ 3 & 0 \end{pmatrix}$

5 $\mathbf{A} = \begin{pmatrix} 3 & 2 \\ x & 6 \end{pmatrix}$ is a singular matrix.

a Show that $x = 9$

b Given that $\mathbf{B} = \begin{pmatrix} 2y & 6 \\ y & 8 \end{pmatrix}$, where y is a constant, is such that $\mathbf{A} + \mathbf{B}$ is a singular matrix, find the value of y.

6 Prove that each of these matrices is singular, where x and y are any real numbers.

a $\mathbf{A} = \begin{pmatrix} 3x & 2x \\ 6y & 4y \end{pmatrix}$ **b** $\mathbf{B} = \begin{pmatrix} x^2 & \sqrt{x} \\ \sqrt{x} & \frac{1}{x} \end{pmatrix}$ **c** $\mathbf{C} = \begin{pmatrix} 1 + \sin x & -\cos x \\ \cos x & \sin x - 1 \end{pmatrix}$

FP1

7 $\mathbf{A} = \begin{pmatrix} -1 & 3 \\ 2 & 4 \end{pmatrix}$ and $\mathbf{B} = \begin{pmatrix} 7 & 4 \\ 3 & k \end{pmatrix}$, where k is a constant.

a Given that det $(\mathbf{AB}) = -20$, show that det $\mathbf{B} = 2$

b Hence find the value of k.

c Find det $(\mathbf{AB}^2)$.

8 Matrix $\mathbf{A} = \begin{pmatrix} p & 1 \\ 2 & q \end{pmatrix}$, where $p < q$ are positive integers,

is such that det $\mathbf{A} = 15$

Find the value of p and the value of q.

9 $\mathbf{P} = \begin{pmatrix} 2 & 1 \\ 6 & 3 \end{pmatrix}$ and $\mathbf{Q} = \begin{pmatrix} -4 & 2 \\ 9 & k \end{pmatrix}$, where k is a constant.

a Given that det $\mathbf{Q} = -4$, find the value of k.

b Show that $\mathbf{PQ} = \frac{1}{2}\mathbf{P}$

c Hence, or otherwise, express $\mathbf{P}^2\mathbf{Q}^2$ in the form $\lambda\mathbf{P}^2$, for λ a constant to be stated.

10 For the matrices $\mathbf{A} = \begin{pmatrix} 2 & 2 \\ 3 & d \end{pmatrix}$ and $\mathbf{B} = \begin{pmatrix} 3 & 2 \\ 5 & 4 \end{pmatrix}$, where d is a constant, it is given that det $(\mathbf{A} + \mathbf{B}) =$ det $\mathbf{A}$ + det $\mathbf{B}$

a Show that $d = \frac{8}{3}$

b Calculate det $(\mathbf{A}^2)$

11 $\mathbf{A} = \begin{pmatrix} x^2+1 & x-1 \\ x+1 & x^2-1 \end{pmatrix}$

a Show that det $\mathbf{A} = x^2(x^2 - 1)$

b Hence find the values of x for which $\mathbf{A}$ is singular.

c Given that det $\mathbf{A} = 12$
find the possible values of the real number x.

12 Given that $\mathbf{A} = \begin{pmatrix} a & b \\ c & d \end{pmatrix}$, use the definition of a determinant to prove that

a det $(\lambda\mathbf{A}) = \lambda^2$ det $\mathbf{A}$, where λ is any real number

b det $(\mathbf{A}^2) =$ (det $\mathbf{A})^2$

5.4 The inverse of a 2 × 2 matrix

The matrix $\mathbf{I} = \begin{pmatrix} 1 & 0 \\ 0 & 1 \end{pmatrix}$ is the 2 × 2 identity matrix.

If **A** is any 2 × 2 matrix then $\mathbf{AI} = \mathbf{A} = \mathbf{IA}$

The matrix **I** acts like the number 1 in ordinary arithmetic.

You can show this using $\mathbf{A} = \begin{pmatrix} a & b \\ c & d \end{pmatrix}$

If **A** is a 2 × 2 non-singular matrix you can always find another 2 × 2 matrix $\mathbf{A}^{-1}$ such that

$\mathbf{AA}^{-1} = \mathbf{I}$ and $\mathbf{A}^{-1}\mathbf{A} = \mathbf{I}$

$\mathbf{A}^{-1}$ is called the inverse matrix of **A**.

A is non-singular if det **A** ≠ 0 – refer to Section 5.3.

If $\mathbf{A} = \begin{pmatrix} a & b \\ c & d \end{pmatrix}$ is non-singular then $\mathbf{A}^{-1} = \frac{1}{\det \mathbf{A}}\begin{pmatrix} d & -b \\ -c & a \end{pmatrix}$

EXAMPLE 1

a Find the inverse of $\mathbf{A} = \begin{pmatrix} 2 & -4 \\ 1 & 3 \end{pmatrix}$

b Verify that $\mathbf{AA}^{-1} = \mathbf{I}$

Only a square matrix can have an inverse.

---

a Compare $\begin{pmatrix} 2 & -4 \\ 1 & 3 \end{pmatrix}$ with $\begin{pmatrix} a & b \\ c & d \end{pmatrix}$

$\det \mathbf{A} = ad - bc = 2 \times 3 - (-4) \times 1 = 10$

Hence $\mathbf{A}^{-1} = \frac{1}{\det \mathbf{A}}\begin{pmatrix} d & -b \\ -c & a \end{pmatrix} = \frac{1}{10}\begin{pmatrix} 3 & 4 \\ -1 & 2 \end{pmatrix}$

You could write this as $\mathbf{A}^{-1} = \begin{pmatrix} 0.3 & 0.4 \\ -0.1 & 0.2 \end{pmatrix}$

b $\mathbf{A} = \begin{pmatrix} 2 & -4 \\ 1 & 3 \end{pmatrix}$, $\mathbf{A}^{-1} = \frac{1}{10}\begin{pmatrix} 3 & 4 \\ -1 & 2 \end{pmatrix}$

$$\mathbf{AA}^{-1} = \frac{1}{10}\begin{pmatrix} 2 & -4 \\ 1 & 3 \end{pmatrix}\begin{pmatrix} 3 & 4 \\ -1 & 2 \end{pmatrix}$$

Bring the constant multiplier $\frac{1}{10}$ to the front of the matrix calculation.

$$= \frac{1}{10}\begin{pmatrix} 2\times3+(-4)\times(-1) & 2\times4+(-4)\times2 \\ 1\times3+3\times(-1) & 1\times4+3\times2 \end{pmatrix}$$

$$= \frac{1}{10}\begin{pmatrix} 10 & 0 \\ 0 & 10 \end{pmatrix} = \begin{pmatrix} 1 & 0 \\ 0 & 1 \end{pmatrix}$$

Multiply each element of $\begin{pmatrix} 10 & 0 \\ 0 & 10 \end{pmatrix}$ by $\frac{1}{10}$.

Hence $\mathbf{AA}^{-1} = \mathbf{I}$, as required.

You can also show that $\mathbf{A}^{-1}\mathbf{A} = \mathbf{I}$

A singular matrix does not have an inverse.

A^{-1} depends on $\frac{1}{\det A}$

If A is singular then $\det A = 0$ and so $\frac{1}{\det A}$ is undefined.

A matrix is a self-inverse if $A^{-1} = A$

EXAMPLE 2

$A = \begin{pmatrix} 3 & -4 \\ k^2 & -3 \end{pmatrix}$, for k any real number.

a Find the possible values of k for which A has no inverse.

b Show that, when $k = \sqrt{2}$, A is self-inverse.

---

a A has no inverse if A is singular

A is singular if det A = 0

$A = \begin{pmatrix} 3 & -4 \\ k^2 & -3 \end{pmatrix}$

$\det A = -9 - (-4k^2)$
$= 4k^2 - 9$

Hence $\det A = 0$ if $4k^2 = 9$

$k = \pm\frac{3}{2}$

Make sure you give both values of k.

The possible values of k are $k = \frac{3}{2}, -\frac{3}{2}$

b If $k = \sqrt{2}$, $A = \begin{pmatrix} 3 & -4 \\ 2 & -3 \end{pmatrix}$

$\det A = -9 - (-8)$
$= -1$

Hence $A^{-1} = \frac{1}{(-1)}\begin{pmatrix} -3 & 4 \\ -2 & 3 \end{pmatrix}$

$= \begin{pmatrix} 3 & -4 \\ 2 & -3 \end{pmatrix}$

$= A$

and so A is self-inverse.

FP1

EXAMPLE 3

If $A = \begin{pmatrix} 1 & 2 \\ -1 & 3 \end{pmatrix}$ and $B = \begin{pmatrix} 2 & -1 \\ 4 & 0 \end{pmatrix}$

show that $(AB)^{-1} = B^{-1}A^{-1}$

First find AB: $AB = \begin{pmatrix} 1 & 2 \\ -1 & 3 \end{pmatrix}\begin{pmatrix} 2 & -1 \\ 4 & 0 \end{pmatrix}$

$= \begin{pmatrix} 10 & -1 \\ 10 & 1 \end{pmatrix}$

Find det AB: $\det AB = 10 - (-10)$

$= 20$

$\therefore\ (AB)^{-1} = \frac{1}{20}\begin{pmatrix} 1 & 1 \\ -10 & 10 \end{pmatrix}$

Find B^{-1}: $\det B = 0 - (-4)$

$= 4$

$B^{-1} = \frac{1}{4}\begin{pmatrix} 0 & 1 \\ -4 & 2 \end{pmatrix}$

Find A^{-1}: $\det A = 3 - (-2)$

$= 5$

$A^{-1} = \frac{1}{5}\begin{pmatrix} 3 & -2 \\ 1 & 1 \end{pmatrix}$

$B^{-1}A^{-1} = \frac{1}{4}\begin{pmatrix} 0 & 1 \\ -4 & 2 \end{pmatrix} \times \frac{1}{5}\begin{pmatrix} 3 & -2 \\ 1 & 1 \end{pmatrix}$

$= \frac{1}{20}\begin{pmatrix} 0 & 1 \\ -4 & 2 \end{pmatrix}\begin{pmatrix} 3 & -2 \\ 1 & 1 \end{pmatrix}$

$= \frac{1}{20}\begin{pmatrix} 1 & 1 \\ -10 & 10 \end{pmatrix}$

$= (AB)^{-1}$

Hence $(AB)^{-1} = B^{-1}A^{-1}$ as required.

If A and B are any pair of non-singular 2×2 matrices then AB is non-singular and its inverse is given by $(AB)^{-1} = B^{-1}A^{-1}$

EXAMPLE 4

Given that AB = C, where A and B are any pair of non-singular 2 × 2 matrices,

a express A in terms of B and C

b simplify the product $CB^{-1}A^{-1}B$

---

a Multiply both sides by B^{-1} on the right:

$$AB = C$$
$$\text{so } ABB^{-1} = CB^{-1}$$
$$\text{i.e. } AI = CB^{-1}$$
$$\text{Hence } A = CB^{-1}$$

$BB^{-1} = I$
$AI = A$

Similarly, starting with AB = C, you can show that $B = A^{-1}C$

b Use $(AB)^{-1} = B^{-1}A^{-1}$:

$$CB^{-1}A^{-1}B = C(B^{-1}A^{-1})B = C(AB)^{-1}B = CC^{-1}B = IB = B$$

Replacing AB with C.

Hence $CB^{-1}A^{-1}B = B$

Exercise 5.4

1 Find the inverse of each matrix.

a $\begin{pmatrix} 4 & 3 \\ 1 & 2 \end{pmatrix}$ **b** $\begin{pmatrix} 6 & 1 \\ -2 & 3 \end{pmatrix}$

c $\begin{pmatrix} 1 & 4 \\ 7 & 3 \end{pmatrix}$ **d** $\begin{pmatrix} \frac{1}{2} & 3 \\ \frac{1}{3} & 4 \end{pmatrix}$

2 $A = \begin{pmatrix} 4 & k \\ k & 2 \end{pmatrix}$, where k is any real number.

a Find A^{-1} in the case when $k = 3$

b Find, in simplified surd form, the values of k for which A has no inverse.

3 **a** Find $\begin{pmatrix} 3 & 2 \\ 2 & 2 \end{pmatrix}^{-1}$

b Hence write down the inverse of the matrix $\begin{pmatrix} 1 & -1 \\ -1 & 1.5 \end{pmatrix}$

FP1

4 Given that $A = \begin{pmatrix} 9 & 2 \\ 3 & -1 \end{pmatrix}$ and $B = \begin{pmatrix} 3 & 4 \\ 7 & 6 \end{pmatrix}$ find, where possible, the inverse of

a A **b** $A + B$ **c** BA

5 $P = \begin{pmatrix} 4 & k \\ 2 & 3 \end{pmatrix}$ where k is a constant, $k \neq 6$

a Find an expression for P^{-1} in terms of k.

b Given that $Q = P^{-1}$ and $q_{21} = \frac{1}{6}$, find the value of k.

6 **a** Prove that the matrix $A = \begin{pmatrix} x & -y \\ y & x \end{pmatrix}$,

where x and y are real numbers, $x \neq 0$, always has an inverse. Write down this inverse in terms of x and y.

b Find A^{-1} in the case when $\det A = 2$, given that x and y are negative integers.

7 **a** Show that the matrix $S = \begin{pmatrix} 4 & -5 \\ 3 & -4 \end{pmatrix}$ is self-inverse.

b Show that there are exactly eight self-inverse matrices of the form $\begin{pmatrix} 4 & c \\ b & -4 \end{pmatrix}$, where b and c are integers.

You need not list these matrices.

8 Let matrix $A = \begin{pmatrix} 6 & 2 \\ 7 & 2 \end{pmatrix}$

a Find A^{-1}.

b Given that $C = AB$ for 2×2 matrices B and C show that $B = A^{-1}C$

c Find B such that $C = \begin{pmatrix} 14 & 19 \\ 16 & 22 \end{pmatrix}$

9 Given that $A = \begin{pmatrix} 3 & 1 \\ 1 & -3 \end{pmatrix}$ and B is a matrix such that

$AB = \begin{pmatrix} 11 & 3 \\ -13 & -4 \end{pmatrix}$

a Find $(AB)^{-1}$.

b Simplify $(AB)^{-1}A$ and hence find B^{-1}.

10 For $\mathbf{P} = \begin{pmatrix} 1 & -2 & 1 \\ 3 & 0 & 2 \end{pmatrix}$ and $\mathbf{Q} = \begin{pmatrix} 4 & 1 \\ -\frac{3}{2} & 0 \\ -6 & -1 \end{pmatrix}$

a show that $\mathbf{PQ} = \mathbf{I}$ where $\mathbf{I}$ is the 2×2 identity matrix.

b Explain briefly why $\mathbf{Q}$ is not the inverse of $\mathbf{P}$.

11 a Show that the simultaneous equations

$$3x + 5y = -8$$
$$2x + 4y = -7$$

can be expressed as the matrix equation

$$\mathbf{A}\begin{pmatrix} x \\ y \end{pmatrix} = \begin{pmatrix} -8 \\ -7 \end{pmatrix} \text{ where } \mathbf{A} = \begin{pmatrix} 3 & 5 \\ 2 & 4 \end{pmatrix}$$

b Find $\mathbf{A}^{-1}$.

c i Show that $\begin{pmatrix} x \\ y \end{pmatrix} = \mathbf{A}^{-1}\begin{pmatrix} -8 \\ -7 \end{pmatrix}$

ii Hence solve the pair of simultaneous equations.

d Use a matrix approach to solve the simultaneous equations

$$2x - 3y = 8$$
$$7x - 5y = 39$$

12 Given that $\mathbf{A} = \begin{pmatrix} a & b \\ c & d \end{pmatrix}$ is non-singular prove,

using appropriate definitions, that

a $\det(\mathbf{A}^{-1}) = \dfrac{1}{\det \mathbf{A}}$

b $(\mathbf{A}^{-1})^{-1} = \mathbf{A}$

13 Given that $\mathbf{A} = \begin{pmatrix} a & b \\ c & d \end{pmatrix}$ has inverse matrix $\mathbf{B} = \begin{pmatrix} e & f \\ g & h \end{pmatrix}$

a Use the equation $\mathbf{AB} = \mathbf{I}$ to show that

$ae + bg = 1$

and $ce + dg = 0$

b Solve this pair of simultaneous equations to show that

$$e = \frac{d}{ad - bc} \quad \text{and} \quad g = \frac{-c}{ad - bc}$$

c Form and solve a pair of simultaneous equations for f and h.

d Hence show that $\mathbf{B} = \dfrac{1}{ad - bc}\begin{pmatrix} d & -b \\ -c & a \end{pmatrix}$

FP1

5.5 Matrices and linear transformations in the x–y plane

You can use a 2×2 matrix to describe a particular type of transformation of the x–y plane.

All transformations in this section refer to the x-y plane only.

Consider the transformation which maps every point $P(x, y)$ to its reflection $P'(-x, y)$ in the y-axis.

$P'(-2, 3)$ is the image of $P(2, 3)$ under this transformation.

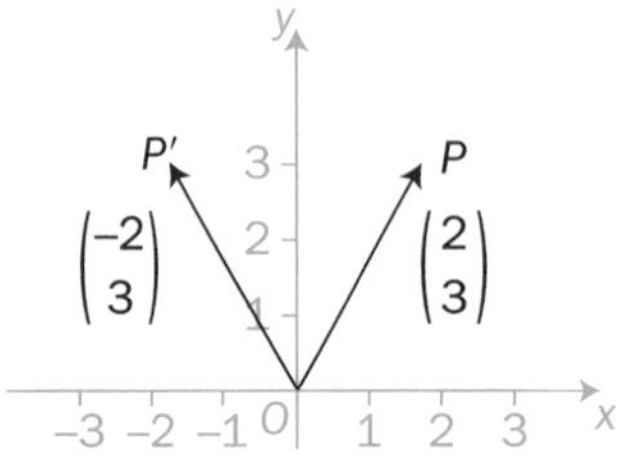

$\begin{pmatrix}-2\\3\end{pmatrix}$ means 'from O, move −2 units parallel to the x-axis and +3 units parallel to the y-axis'.

Because of the way matrix multiplication is defined, it is convenient to represent the points $P(2, 3)$ and $P'(-2, 3)$ by the vectors $\begin{pmatrix}2\\3\end{pmatrix}$ and $\begin{pmatrix}-2\\3\end{pmatrix}$ respectively.

Under this transformation, $\begin{pmatrix}2\\3\end{pmatrix}$ is mapped onto $\begin{pmatrix}-2\\3\end{pmatrix}$

You can write this as $\begin{pmatrix}2\\3\end{pmatrix} \to \begin{pmatrix}-2\\3\end{pmatrix}$

The effect of this transformation on $\begin{pmatrix}2\\3\end{pmatrix}$ can also be described by

multiplying the 2×1 matrix $\begin{pmatrix}2\\3\end{pmatrix}$ by the 2×2 matrix $\mathbf{A} = \begin{pmatrix}-1 & 0\\0 & 1\end{pmatrix}$

$$\mathbf{A}\begin{pmatrix}2\\3\end{pmatrix} = \begin{pmatrix}-1 & 0\\0 & 1\end{pmatrix}\begin{pmatrix}2\\3\end{pmatrix}$$
$$= \begin{pmatrix}-1\times 2 + 0\times 3\\0\times 2 + 1\times 3\end{pmatrix} = \begin{pmatrix}-2\\3\end{pmatrix}$$

Because $\mathbf{A}$ has order 2×2, the product $\mathbf{A}\begin{pmatrix}2\\3\end{pmatrix}$ exists and is a 2×1 matrix – refer to Section 5.2.

In general, $\mathbf{A}\begin{pmatrix}x\\y\end{pmatrix} = \begin{pmatrix}-1 & 0\\0 & 1\end{pmatrix}\begin{pmatrix}x\\y\end{pmatrix}$

$$= \begin{pmatrix}-1\times x + 0\times y\\0\times x + 1\times y\end{pmatrix}$$
$$= \begin{pmatrix}-x\\y\end{pmatrix}$$

$\mathbf{A}\begin{pmatrix}x\\y\end{pmatrix} = \begin{pmatrix}-x\\y\end{pmatrix}$ is equivalent to the transformation $\begin{pmatrix}x\\y\end{pmatrix} \to \begin{pmatrix}-x\\y\end{pmatrix}$

FP1

You can therefore represent a reflection in the y-axis by the matrix $\begin{pmatrix} -1 & 0 \\ 0 & 1 \end{pmatrix}$

EXAMPLE 1

Find the matrix which represents a reflection in the x-axis.

Let $\mathbf{A} = \begin{pmatrix} a & b \\ c & d \end{pmatrix}$ be the matrix which represents a reflection in the x-axis.

Then $\mathbf{A}\begin{pmatrix} x \\ y \end{pmatrix} = \begin{pmatrix} x \\ -y \end{pmatrix}$ for all possible values of x and y.

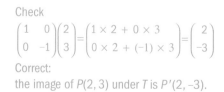

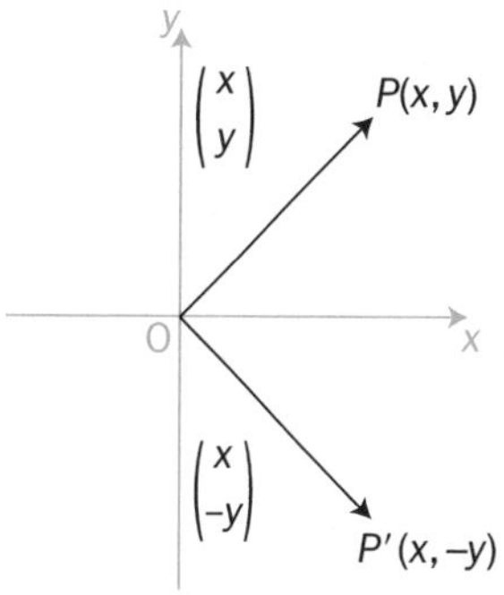

So $\begin{pmatrix} a & b \\ c & d \end{pmatrix}\begin{pmatrix} x \\ y \end{pmatrix} = \begin{pmatrix} x \\ -y \end{pmatrix}$ (1)

Equation (1) must hold for *all* values of x and y.
In particular you can choose $x = 1$ and $y = 0$

The aim is to find the values of a, b, c and d.

Substitute $x = 1$, $y = 0$ into (1):

$$\begin{pmatrix} a & b \\ c & d \end{pmatrix}\begin{pmatrix} 1 \\ 0 \end{pmatrix} = \begin{pmatrix} 1 \\ 0 \end{pmatrix}$$

$$\begin{pmatrix} a\times 1 + b\times 0 \\ c\times 1 + d\times 0 \end{pmatrix} = \begin{pmatrix} 1 \\ 0 \end{pmatrix}$$

$$\begin{pmatrix} a \\ c \end{pmatrix} = \begin{pmatrix} 1 \\ 0 \end{pmatrix}$$

For any 2 × 2 matrix **A**,

$\mathbf{A}\begin{pmatrix} 1 \\ 0 \end{pmatrix}$ = 1st column of **A**.

Hence $a = 1$ and $c = 0$

Similarly, substitute $x = 0$ and $y = 1$ into (1) to find b and d:

$$\begin{pmatrix} a & b \\ c & d \end{pmatrix}\begin{pmatrix} 0 \\ 1 \end{pmatrix} = \begin{pmatrix} 0 \\ -1 \end{pmatrix}$$

$$\begin{pmatrix} a\times 0 + b\times 1 \\ c\times 0 + d\times 1 \end{pmatrix} = \begin{pmatrix} 0 \\ -1 \end{pmatrix}$$

$$\begin{pmatrix} b \\ d \end{pmatrix} = \begin{pmatrix} 0 \\ -1 \end{pmatrix}$$

For any 2 × 2 matrix **A**,

$\mathbf{A}\begin{pmatrix} 0 \\ 1 \end{pmatrix}$ = 2nd column of **A**

Hence $b = 0$ and $d = -1$

The matrix $\mathbf{A} = \begin{pmatrix} 1 & 0 \\ 0 & -1 \end{pmatrix}$

Check

$\begin{pmatrix} 1 & 0 \\ 0 & -1 \end{pmatrix}\begin{pmatrix} 2 \\ 3 \end{pmatrix} = \begin{pmatrix} 1\times 2 + 0\times 3 \\ 0\times 2 + (-1)\times 3 \end{pmatrix} = \begin{pmatrix} 2 \\ -3 \end{pmatrix}$

Correct:
the image of $P(2, 3)$ under T is $P'(2, -3)$.

FP1

If the matrix $A = \begin{pmatrix} a & b \\ c & d \end{pmatrix}$ represents a transformation then the first and second columns of A are found by applying the transformation to the vectors $\begin{pmatrix} 1 \\ 0 \end{pmatrix}$ and $\begin{pmatrix} 0 \\ 1 \end{pmatrix}$ respectively, that is

$$\begin{pmatrix} 1 \\ 0 \end{pmatrix} \to \begin{pmatrix} a \\ c \end{pmatrix}, \begin{pmatrix} 0 \\ 1 \end{pmatrix} \to \begin{pmatrix} b \\ d \end{pmatrix}$$

EXAMPLE 2

Find the matrix A which represents a 90° anticlockwise rotation, centre the origin.

Let A be the matrix that represents a rotation of 90° anticlockwise, centre the origin. In this transformation

$$\begin{pmatrix} 1 \\ 0 \end{pmatrix} \to \begin{pmatrix} 0 \\ 1 \end{pmatrix}$$

and, similarly, $\begin{pmatrix} 0 \\ 1 \end{pmatrix} \to \begin{pmatrix} -1 \\ 0 \end{pmatrix}$

Hence the matrix A has first and second columns $\begin{pmatrix} 0 \\ 1 \end{pmatrix}$ and $\begin{pmatrix} -1 \\ 0 \end{pmatrix}$ respectively.

i.e. $A = \begin{pmatrix} 0 & -1 \\ 1 & 0 \end{pmatrix}$

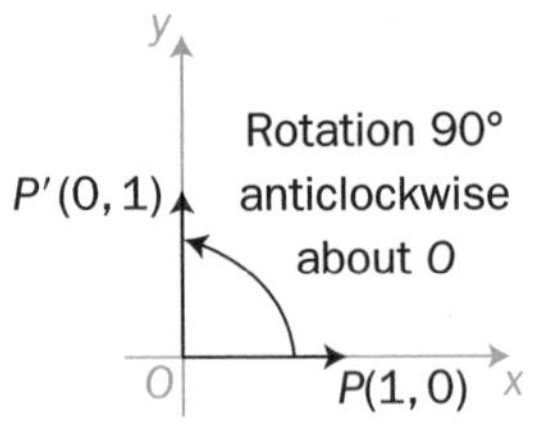

A transformation is linear if it can be represented by a matrix.

e.g. A rotation of 180°, about the point (1, 1), is a non-linear transformation – no matrix can be found to represent this transformation.

See question 4 of Exercise 5.5.

You need to be familiar with the following linear transformations and their matrix representations.

FP1

Transformation	Matrix representation
Reflection in the y-axis	$\begin{pmatrix} -1 & 0 \\ 0 & 1 \end{pmatrix}$
Reflection in the x-axis	$\begin{pmatrix} 1 & 0 \\ 0 & -1 \end{pmatrix}$
Reflection in the line $y = x$	$\begin{pmatrix} 0 & 1 \\ 1 & 0 \end{pmatrix}$
Reflection in the line $y = -x$	$\begin{pmatrix} 0 & -1 \\ -1 & 0 \end{pmatrix}$
Anticlockwise rotation through angle θ, about O	$\begin{pmatrix} \cos\theta & -\sin\theta \\ \sin\theta & \cos\theta \end{pmatrix}$
Enlargement scale factor k centre O	$\begin{pmatrix} k & 0 \\ 0 & k \end{pmatrix}$

You can find each of these matrices by considering the effect each transformation has on the vectors $\begin{pmatrix} 1 \\ 0 \end{pmatrix}$ and $\begin{pmatrix} 0 \\ 1 \end{pmatrix}$.

This result is in the formula book. By choosing appropriate values of θ you can write down the matrices for anticlockwise rotations through 90°, 180° and 270°.

FP1

EXAMPLE 3

The matrix $\mathbf{A} = -\frac{1}{\sqrt{2}}\begin{pmatrix} 1 & -1 \\ 1 & 1 \end{pmatrix}$ represents an anticlockwise rotation of θ, centre the origin, where $0° \leqslant \theta \leqslant 360°$.
Find θ.

Let $\mathbf{A} = \begin{pmatrix} \cos\theta & -\sin\theta \\ \sin\theta & \cos\theta \end{pmatrix} = -\frac{1}{\sqrt{2}}\begin{pmatrix} 1 & -1 \\ 1 & 1 \end{pmatrix}$

$$= \begin{pmatrix} -\frac{1}{\sqrt{2}} & \frac{1}{\sqrt{2}} \\ -\frac{1}{\sqrt{2}} & -\frac{1}{\sqrt{2}} \end{pmatrix}$$

$\cos\theta = -\frac{1}{\sqrt{2}}$ and $\sin\theta = -\frac{1}{\sqrt{2}}$

Hence $\tan\theta = 1$

$$\tan\theta \equiv \frac{\sin\theta}{\cos\theta} = \frac{\left(-\frac{1}{\sqrt{2}}\right)}{\left(-\frac{1}{\sqrt{2}}\right)} = 1$$

$\therefore\ \theta = 45°, 225°$ for $0° \leqslant \theta \leqslant 360°$

Refer to C2.

$\cos\theta < 0$ and $\sin\theta < 0$, so θ lies in third quadrant
Hence the angle of rotation $\theta = 225°$

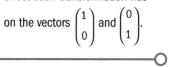

Exercise 5.5

1 By considering the effect each transformation has on the vectors $\begin{pmatrix}1\\0\end{pmatrix}$ and $\begin{pmatrix}0\\1\end{pmatrix}$, find the matrix which represents the given transformation.

a Rotation of 180°, centre the origin

b Reflection in the line $y = x$

c Rotation of 270° anticlockwise, about the origin

d Reflection in the line $y = -x$

e Enlargement, scale factor 1.5, about the origin

2 For a rotation of 45° anticlockwise, about the origin

a show that the vector $\begin{pmatrix}1\\0\end{pmatrix}$ is mapped onto the vector $\begin{pmatrix}\frac{1}{\sqrt{2}}\\\frac{1}{\sqrt{2}}\end{pmatrix}$

Draw a diagram.

b find the vector onto which the vector $\begin{pmatrix}0\\1\end{pmatrix}$ is mapped

c hence write down the matrix which represents this transformation.

3 An enlargement, scale factor k, centre the origin, maps the point (2, 6) to the point (7, 21).

a Find the value of k.

b Write down the matrix which represents this enlargement.

4 A transformation is represented by the matrix $\mathrm{A} = \begin{pmatrix}a & b\\c & d\end{pmatrix}$

a By considering the matrix product $\mathrm{A}\begin{pmatrix}0\\0\end{pmatrix}$, show that the transformation maps the origin onto itself.

b Hence explain why a rotation of 180°, about (1, 1), cannot be represented by a matrix.

FP1

5 In the diagram, the lines OA and OB are the result of rotating clockwise through an angle θ, about O, the vectors $\begin{pmatrix}1\\0\end{pmatrix}$ and $\begin{pmatrix}0\\1\end{pmatrix}$ respectively.

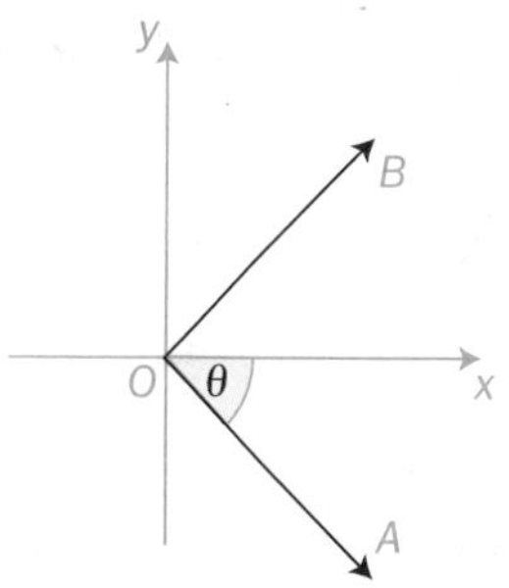

a i Show that point A has coordinates $(\cos\theta, -\sin\theta)$.

ii Find a similar expression for the coordinates of point B.

b Hence write down the matrix **P** which represents a clockwise rotation of angle θ, about the origin.

c Show that det $\mathbf{P} = 1$.

6 It is given that an anticlockwise rotation of an angle θ about the origin is represented by the matrix $\begin{pmatrix}\cos\theta & -\sin\theta\\ \sin\theta & \cos\theta\end{pmatrix}$

a Find the matrix which represents an anticlockwise rotation of 135°, about the origin. Give the elements of this matrix in simplified surd form.

An anticlockwise rotation through an angle θ, about the origin, is represented by the matrix $\frac{1}{\sqrt{2}}\begin{pmatrix}1 & 1\\ -1 & 1\end{pmatrix}$, where $0° \leqslant \theta \leqslant 360°$.

b Find θ.

7 The diagram shows the line with equation $y = \sqrt{3}x$

The line makes an acute angle θ with the positive x-axis.

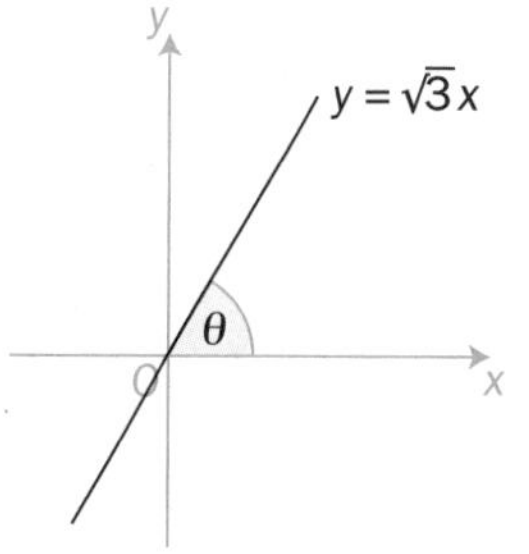

a Show that $\theta = 60°$

b Find the matrix which represents a reflection in the line $y = \sqrt{3}x$

c Show in a similar way that the matrix which represents a reflection in the line $\sqrt{3}y = x$ is given by $\frac{1}{2}\begin{pmatrix}1 & \sqrt{3}\\ \sqrt{3} & -1\end{pmatrix}$

d Show that the matrix which represents a reflection in the line $y = x\tan\theta$ for any acute angle θ, is given by $\begin{pmatrix}\cos 2\theta & \sin 2\theta\\ \sin 2\theta & -\cos 2\theta\end{pmatrix}$

FP1

5.6 Combined and inverse transformations

You can use matrix multiplication to represent one transformation followed by another.

If two transformations are represented by matrices **A** and **B** respectively then **BA** is the matrix which represents transformation **A** followed by transformation **B**.

BA means **A** first then **B**.

EXAMPLE 1

Find the matrix which represents the combined transformation of a rotation of 180°, about the origin, followed by a reflection in the line $y = x$
Comment on your answer.

The matrix which represents the given rotation is

$$\mathbf{A} = \begin{pmatrix} -1 & 0 \\ 0 & -1 \end{pmatrix}$$

Refer to Section 5.5.

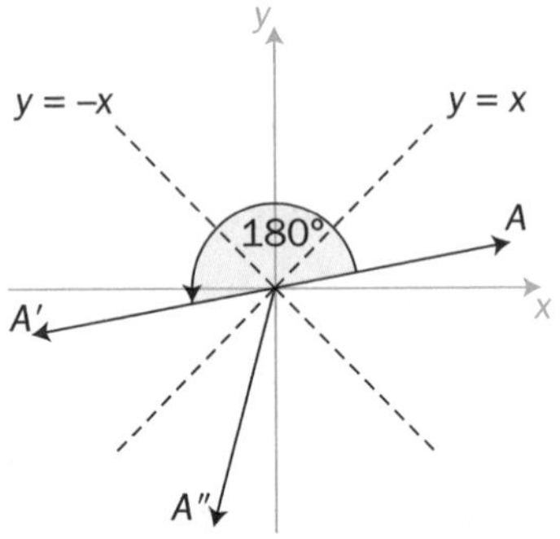

The matrix which represents the given reflection

is $\mathbf{B} = \begin{pmatrix} 0 & 1 \\ 1 & 0 \end{pmatrix}$

Hence the matrix which represents the combined transformation of the rotation followed by the reflection is the product **BA**.

$$\mathbf{BA} = \begin{pmatrix} 0 & 1 \\ 1 & 0 \end{pmatrix}\begin{pmatrix} -1 & 0 \\ 0 & -1 \end{pmatrix} = \begin{pmatrix} (0\times -1)+(1\times 0) & (0\times 0)+(1\times -1) \\ (1\times -1)+(0\times 0) & (1\times 0)+(0\times -1) \end{pmatrix}$$

$$= \begin{pmatrix} 0 & -1 \\ -1 & 0 \end{pmatrix}$$

BA is the transformation represented by **A** *followed by* that represented by **B**.

The matrix **BA** represents a reflection in the line $y = -x$

Refer to Section 5.5.

Hence a rotation of 180°, about the origin, followed by a reflection in the line $y = x$ is equivalent to a reflection in the line $y = -x$

Two transformations can sometimes be replaced by a single transformation.

FP1

Given one transformation you can sometimes find another transformation which reverses its effect.
When this happens, the two transformations are inverses of each other.

Any point is left unchanged as a result of these two transformations.

e.g. The inverse of an enlargement, about O, scale factor 2 is an enlargement, about O, scale factor $\frac{1}{2}$, and vice versa.

If a transformation is represented by matrix A then the inverse transformation (if it exists) is represented by $\mathbf{A}^{-1}$.

The inverse of **A** exists provided that det $\mathbf{A} \neq 0$ – refer to Section 5.4.

EXAMPLE 2

a Write down the matrix A which represents an anticlockwise rotation through angle θ, about the origin.

b Hence find the matrix which represents a clockwise rotation through angle θ, about the origin.

---

a $\mathbf{A} = \begin{pmatrix} \cos\theta & -\sin\theta \\ \sin\theta & \cos\theta \end{pmatrix}$

This matrix is given in the formula book.

b Since a clockwise rotation through angle θ, about the origin, is the inverse transformation it is represented by the matrix $\mathbf{A}^{-1}$.

Find det **A** and then $\mathbf{A}^{-1}$:

If $\mathbf{A} = \begin{pmatrix} \cos\theta & -\sin\theta \\ \sin\theta & \cos\theta \end{pmatrix}$,

$$\det \mathbf{A} = \cos^2\theta - (-\sin^2\theta)$$

Use $\left|\begin{pmatrix} a & b \\ c & d \end{pmatrix}\right| = ad - bc$

Refer to Section 5.3.

$$= \cos^2\theta + \sin^2\theta$$
$$= 1$$

$\cos^2\theta + \sin^2\theta = 1$ Refer to C2.

$$\therefore \mathbf{A}^{-1} = \frac{1}{1}\begin{pmatrix} \cos\theta & \sin\theta \\ -\sin\theta & \cos\theta \end{pmatrix}$$

If $\mathbf{A} = \begin{pmatrix} a & b \\ c & d \end{pmatrix}$

$\mathbf{A}^{-1} = \frac{1}{\det \mathbf{A}}\begin{pmatrix} d & -b \\ -c & a \end{pmatrix}$

Refer to Section 5.4.

Hence a clockwise rotation through angle θ, about the origin, is represented by the matrix $\begin{pmatrix} \cos\theta & \sin\theta \\ -\sin\theta & \cos\theta \end{pmatrix}$

FP1

The inverse of the combined transformation represented by BA is the transformation represented by $(\mathbf{BA})^{-1}$, that is $\mathbf{A}^{-1}\mathbf{B}^{-1}$.

BA means that transformation **A** is done first, then transformation **B**.

EXAMPLE 3

Two transformations are represented by $\mathbf{A} = \begin{pmatrix} 0 & -1 \\ 1 & 0 \end{pmatrix}$ and $\mathbf{B} = \begin{pmatrix} 0 & 1 \\ 1 & 0 \end{pmatrix}$.

Find the matrix which represents the inverse of the combined transformation represented by BA.
Comment on your answer.

The matrix which represents the inverse of the combined transformation BA is $\mathbf{A}^{-1}\mathbf{B}^{-1}$.

Refer to Section 5.4.

$$\mathbf{A} = \begin{pmatrix} 0 & -1 \\ 1 & 0 \end{pmatrix} \text{ and } \mathbf{A}^{-1} = \begin{pmatrix} 0 & 1 \\ -1 & 0 \end{pmatrix}$$

$\det \mathbf{A} = (0 \times 0) - (1 \times -1) = 1$

$$\mathbf{B} = \begin{pmatrix} 0 & 1 \\ 1 & 0 \end{pmatrix} \text{ and } \mathbf{B}^{-1} = \frac{1}{-1}\begin{pmatrix} 0 & -1 \\ -1 & 0 \end{pmatrix}$$

$$= \begin{pmatrix} 0 & 1 \\ 1 & 0 \end{pmatrix}$$

$\det \mathbf{B} = (0 \times 0) - (1 \times 1) = -1$

Notice that **B** is self-inverse.

$$\text{Hence } \mathbf{A}^{-1}\mathbf{B}^{-1} = \begin{pmatrix} 0 & 1 \\ -1 & 0 \end{pmatrix}\begin{pmatrix} 0 & 1 \\ 1 & 0 \end{pmatrix}$$

$$= \begin{pmatrix} 1 & 0 \\ 0 & -1 \end{pmatrix}$$

Inspect the columns of this matrix:

The inverse of the combined transformation BA is a reflection in the x-axis.

FP1

You can interpret the determinant of a matrix as an area scale factor.

e.g. You can see the effect the matrix $\mathbf{M} = \begin{pmatrix} 3 & 0 \\ 0 & 2 \end{pmatrix}$ has on the x–y plane by looking at the square $ABCD$, where the vertices have coordinates $A(1, 1)$, $B(2, 1)$, $C(2, 2)$ and $D(1, 2)$.

Since $\begin{pmatrix} 3 & 0 \\ 0 & 2 \end{pmatrix}\begin{pmatrix} 1 \\ 1 \end{pmatrix} = \begin{pmatrix} 3 \\ 2 \end{pmatrix}$, $\mathbf{M}$ maps point $A(1, 1)$ onto point $A'(3, 2)$.

The effect of $\mathbf{M}$ on the vertices B, C and D can be found in the same way.

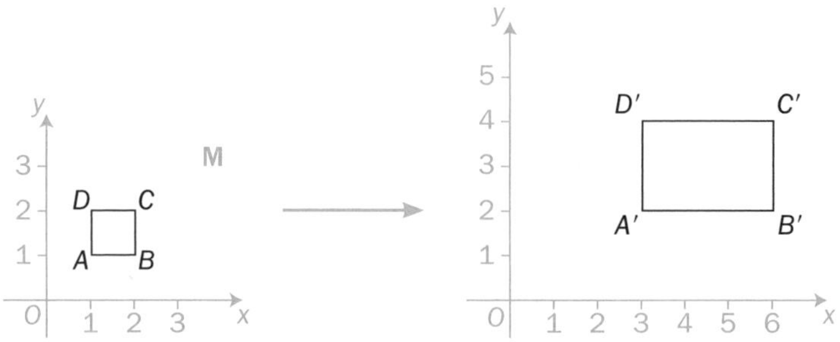

$A'B'C'D'$ is the image of $ABCD$ under matrix $\mathbf{M}$.

$\mathbf{M}$ maps the square $ABCD$, of area 1, onto rectangle $A'B'C'D'$, which has area 6.

The area scale factor = 6

$\det \mathbf{M} = (3 \times 2) - (0 \times 0) = 6$

So the area scale factor is equal to the determinant of $\mathbf{M}$.

The matrix $\begin{pmatrix} a & b \\ c & d \end{pmatrix}$ maps any shape of area k onto one with area $(ad - bc)k$

$\left|\begin{pmatrix} a & b \\ c & d \end{pmatrix}\right| = ad - bc$ is the area scale factor by which $\begin{pmatrix} a & b \\ c & d \end{pmatrix}$ transforms the x–y plane.

You need to modify this result when $ad - bc < 0$.

If the matrix M has a negative determinant, the transformation represented by M involves a reflection.

FP1

EXAMPLE 4

A triangle in the x–y plane has area 5 square units.
Find the area of the image of this triangle when it is transformed by the matrix $\begin{pmatrix} 4 & 2 \\ 1 & 3 \end{pmatrix}$

Work out the determinant of the transforming matrix:

$$\left|\begin{pmatrix} 4 & 2 \\ 1 & 3 \end{pmatrix}\right| = (4 \times 3) - (1 \times 2) = 10$$

The area scale factor is 10.

Since the given triangle has area 5, its image has area $10 \times 5 = 50$ square units.

EXAMPLE 5

The square with vertices O, $A(1, 0)$, $B(1, 1)$, $C(0, 1)$ is mapped onto the quadrilateral $OA'B'C'$ by the matrix $\mathbf{M} = \begin{pmatrix} 4 & 5 \\ 2 & 1 \end{pmatrix}$

a Illustrate $OA'B'C'$ and describe its shape.

b Calculate det **M**. Comment on your answer.

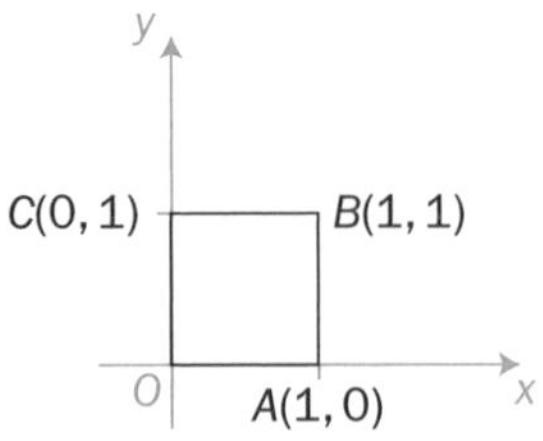

a The origin O is always mapped to itself by any matrix.

$OABC$ is a unit square.

Work out the coordinates of A', B' and C':

$\begin{pmatrix} 4 & 5 \\ 2 & 1 \end{pmatrix}\begin{pmatrix} 1 \\ 0 \end{pmatrix} = \begin{pmatrix} 4 \\ 2 \end{pmatrix}$ so A' has coordinates (4, 2)

$\begin{pmatrix} 4 & 5 \\ 2 & 1 \end{pmatrix}\begin{pmatrix} 1 \\ 1 \end{pmatrix} = \begin{pmatrix} 9 \\ 3 \end{pmatrix}$ so B' has coordinates (9, 3)

$\begin{pmatrix} 4 & 5 \\ 2 & 1 \end{pmatrix}\begin{pmatrix} 0 \\ 1 \end{pmatrix} = \begin{pmatrix} 5 \\ 1 \end{pmatrix}$ so C' has coordinates (5, 1)

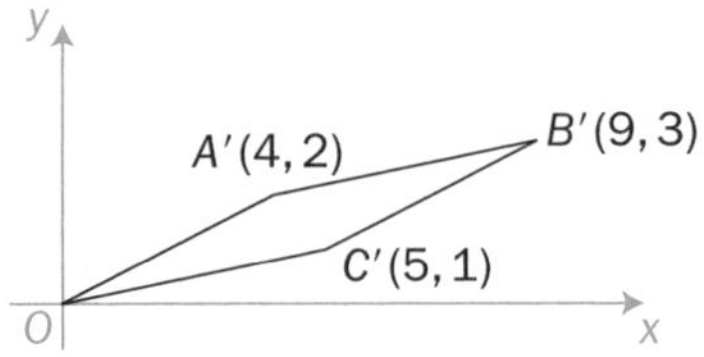

$OA'B'C'$ is a parallelogram.

e.g. By using their coordinates, you can see that $B'C'$ is equal in length and direction to the line OA'.

b $\det \mathbf{M} = \left|\begin{pmatrix} 4 & 5 \\ 2 & 1 \end{pmatrix}\right| = (4 \times 1) - (2 \times 5) = -6$

The area of $OA'B'C'$ is 6.

The negative sign indicates that, as well as changing its shape, the square $OABC$ has been turned over (or reflected).

The image is a reflection of the original square in the diagonal OB.

FP1

Exercise 5.6

1 For each of the following

i write down the matrices **A** and **B** which represent the two transformations

ii find the matrix which represents the combined transformation of **A** followed by **B**.

a **A**: A rotation of 180°, about the origin

B: A reflection in the y-axis

b **A**: A reflection in the x-axis

B: An anticlockwise rotation of 90°, about the origin

c **A**: An enlargement, scale factor 5, about the origin

B: A reflection in the line $y = -x$

d **A**: An anticlockwise rotation of 45°, about the origin

B: A reflection in the line $y = x$

2 Matrices **A** and **B** represent reflections in the y-axis and in the line $y = -x$ respectively.

a Write down **A** and **B**.

b Hence find the matrix **C** which represents the combined transformation of a reflection in the line $y = -x$ followed by a reflection in the y-axis.

c State a single transformation represented by the matrix **C**.

3 Given that the matrices **P** and **Q** represent, respectively, anticlockwise rotations of 45° and 90°, about the origin,

a write down **P** and **Q**

b give a geometrical reason why $\mathbf{P}^2 = \mathbf{Q}$

c verify that $\mathbf{P}^2 = \mathbf{Q}$ by calculation

d simplify $\mathbf{P}^8$.

4 Matrices **P** and **Q** represent reflections in the line $y = x$ and the x-axis respectively.

a Find **P** and **Q**.

b Find **PQ** and describe a single transformation that this product represents.

c Hence find the smallest positive integer n for which $(\mathbf{PQ})^{2n} = \mathbf{I}$, where **I** is the 2×2 identity matrix.

FP1

5 $\mathbf{R} = \begin{pmatrix} 0 & 1 \\ -1 & 0 \end{pmatrix}$

a By considering its columns, write down a single transformation represented by **R**.

b Hence write down the transformation represented by $\mathbf{R}^{-1}$.

6 a Give a geometrical reason why the matrix **A** which represents a reflection in the line $y = x$ is self-inverse.

b Write down **A** and verify that it is a self-inverse matrix.

A matrix **A** is self-inverse if $\mathbf{A}^{-1} = \mathbf{A}$ – refer to Section 5.4.

7 a Find the matrix **P** which represents the combined transformation of an anticlockwise rotation of 135° about the origin, followed by an enlargement, scale factor $\sqrt{2}$, centre the origin.

b Show that $\mathbf{P}^{-1} = \frac{1}{2}\begin{pmatrix} -1 & 1 \\ -1 & -1 \end{pmatrix}$

c State, in the correct order, two transformations whose combination is represented by $\mathbf{P}^{-1}$.

8 Find the area scale factor for

a the transformation represented by the matrix $\mathbf{A} = \begin{pmatrix} 3 & 2 \\ 6 & 5 \end{pmatrix}$

b the transformation represented by the matrix $\mathbf{B} = \begin{pmatrix} 7 & 6 \\ 3 & 2 \end{pmatrix}$

c a reflection in the line $y = x$

d an anticlockwise rotation through θ, about the origin.

9 Under matrix $\mathbf{P} = \begin{pmatrix} 5 & 1 \\ 7 & 3 \end{pmatrix}$, a triangle ABC is mapped onto triangle $A'B'C'$.

a Given that triangle ABC has area 10 square units, find the area of triangle $A'B'C'$.

b Find the area of the shape which maps onto triangle ABC under **P**.

FP1

10 Under matrix $\mathbf{M} = \begin{pmatrix} 10 & 4 \\ 6 & 3 \end{pmatrix}$, a square S of area 4 square units is transformed into a parallelogram P.

a Calculate the area of P.

b Given that a pair of adjacent sides of P have lengths 6 and 8 units, find the two possible angles between this pair of sides.

11 A transformation is represented by the matrix $\mathbf{M} = \begin{pmatrix} 2 & 3 \\ 4 & 6 \end{pmatrix}$
The diagram shows a unit square.

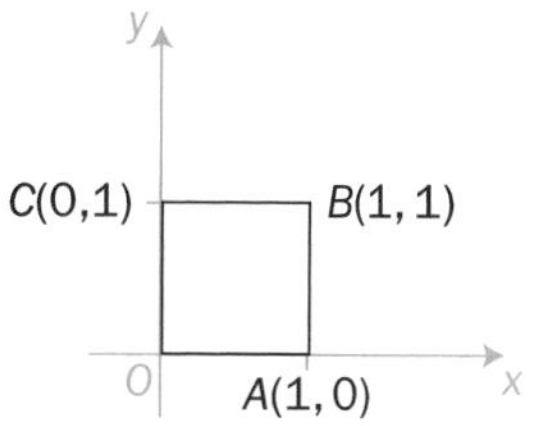

a Show that $\mathbf{M}$ maps points A, B and C onto points which all lie on the straight line with equation $y = 2x$

b Explain how this result is consistent with the result that det $\mathbf{M}$ is the area scale factor of the transformation represented by $\mathbf{M}$.

12 The matrix $\mathbf{Q}$ transforms a shape S of area 4 square units into a shape with area 18 square units.

a Write down the two possible values of det $\mathbf{Q}$.

b Given that $\mathbf{Q} = \begin{pmatrix} 6 & 5 \\ 1.5 & x \end{pmatrix}$, find the possible values of x.

c Find the area of the shape which is the image of S under $\mathbf{Q}^{-1}$.

13 The notation $\mathbf{R}(\theta)$ refers to the matrix representing an anticlockwise rotation through angle θ, about the origin.

a Write down $\mathbf{R}(\theta)$.

b Show that $\mathbf{R}^2(\theta) = \begin{pmatrix} \cos^2\theta - \sin^2\theta & -2\sin\theta\cos\theta \\ 2\sin\theta\cos\theta & \cos^2\theta - \sin^2\theta \end{pmatrix}$

c i Deduce that $\cos 2\theta \equiv \cos^2\theta - \sin^2\theta$ and find an identity for $\sin 2\theta$.
ii Hence find an identity for $\tan 2\theta$ in terms of $\tan\theta$.

d By considering the matrix $\mathbf{R}(3\theta)$, show that $\cos 3\theta \equiv 4\cos^3\theta - 3\cos\theta$

FP1

Review 5

1 $\mathbf{A} = \begin{pmatrix} 3 & -2 & 4 \\ 1 & 5 & -1 \end{pmatrix}$, $\mathbf{B} = \begin{pmatrix} 2 & 6 \\ 0 & -2 \\ -3 & 1 \end{pmatrix}$

a Show that $\mathbf{AB} = \begin{pmatrix} -6 & 26 \\ 5 & -5 \end{pmatrix}$

b Find $(\mathbf{AB})^{-1}$.

c Explain briefly why the result $(\mathbf{AB})^{-1} = \mathbf{B}^{-1}\mathbf{A}^{-1}$ does **not** apply in this case.

2 a Given the matrix $\mathbf{A} = \begin{pmatrix} 4 & -3 \\ 2 & 1 \end{pmatrix}$ find the constants s and t

such that $\mathbf{A}^2 = s\mathbf{A} + t\mathbf{I}$

where $\mathbf{I}$ is the 2×2 identity matrix.

b Hence, or otherwise, express $\mathbf{A}^3$ in the form $p\mathbf{A} + q\mathbf{I}$ for constants p and q.

FP1

3 Given that $\mathbf{A} = \begin{pmatrix} 2 & -4 \\ 3 & -2 \end{pmatrix}$ and $\mathbf{B}$ is a matrix such that $\mathbf{AB} = 2\begin{pmatrix} 0 & 4 \\ 1 & 4 \end{pmatrix}$

a find $\mathbf{A}^{-1}$

b hence, or otherwise, show that $\mathbf{B} = \begin{pmatrix} 1 & 2 \\ \frac{1}{2} & -1 \end{pmatrix}$

c find $\mathbf{B}^2$ and hence write down $\mathbf{B}^{-1}$.

4 For the 2×2 non-singular matrices $\mathbf{A}$ and $\mathbf{B}$, simplify

a $\mathbf{A}(\mathbf{BA})^{-1}\mathbf{B}$

b $(\mathbf{A}^{-1}\mathbf{B}^{-1})^{-1}$

c $\mathbf{B}(\mathbf{A}^{-1}\mathbf{B})^{-1}$

5 $\mathbf{G} = \begin{pmatrix} 1 & 1 \\ -1 & -1 \end{pmatrix}$ and $\mathbf{H} = \begin{pmatrix} p & q \\ q & p \end{pmatrix}$, where p and q are any real numbers.

a Show that $\mathbf{G}$ is singular and find an expression for det $\mathbf{H}$.

b Verify that det $(\mathbf{G} + \mathbf{H}) = \det \mathbf{G} + \det \mathbf{H}$.

Take care. This rule is **not** generally true.

c Find an expression for $\mathbf{GH}$ and hence show that $\mathbf{GH}$ is singular.

6 a Write down

i the matrix **A** which represents a reflection in the y-axis
ii the matrix **B** which represents a rotation of 180° about the origin
iii the matrix **C** which represents a reflection in the line $y = -x$

b By using matrix methods find

i the image of the point (3, 2) after a rotation of 180° about the origin followed by a reflection in the y-axis
ii the coordinates of the point whose image is the point (4, −5) after a reflection in the line $y = -x$

7 a Write down the matrix **M** which represents a 90° anticlockwise rotation, about the origin.

b State the transformation represented by the matrix $\mathbf{N} = \begin{pmatrix} 0 & 1 \\ 1 & 0 \end{pmatrix}$ and show that **N** is its own inverse.

c Find **MN** and state the single transformation this matrix represents.

d Hence, or otherwise, find $\mathbf{NM}^{-1}$

8 The diagram shows a triangle OAB where the vertices A and B have coordinates $A(0, 5)$ and $B(k, 2)$.

Matrix $\mathbf{M} = \begin{pmatrix} 4 & 2 \\ 3 & 3 \end{pmatrix}$ transforms triangle OAB into one with area 45 square units.

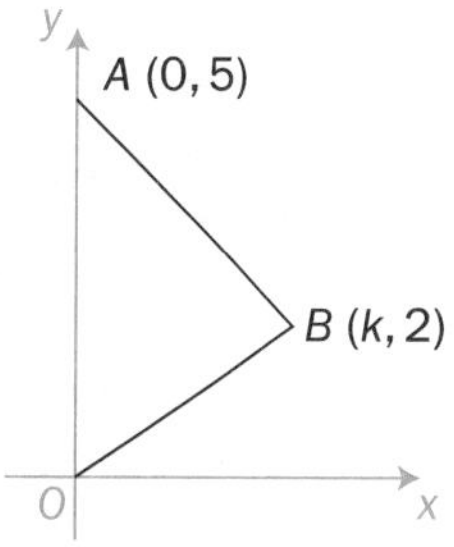

a Find det **M** and hence show that the area of triangle OAB is 7.5 square units.

b Find the value of k and hence find the coordinates of the image of B under **M**.

9 $\mathbf{P} = \begin{pmatrix} \cos\alpha & \sin\alpha \\ \sin\alpha & \cos\alpha \end{pmatrix}$ where α is an acute angle such that **P** is singular.

a Find the value of α

b Show that $\mathbf{P}^2 = \sqrt{2}\mathbf{P}$

c Write down the matrix **Q** which represents an anticlockwise rotation through α, about the origin.

d Express $\mathbf{Q}^4\mathbf{P}^4$ in the form $\lambda\mathbf{P}$, where the exact value of λ is to be stated.

10 a Find the value of x for which $\mathbf{A} = \begin{pmatrix} 9 & 1 \\ 4 & x^2 \end{pmatrix}$ is singular and for which $\mathbf{B} = \begin{pmatrix} 4 & 8 \\ 1 & 3x \end{pmatrix}$ is non-singular.

b Show that $\begin{pmatrix} \cos^2\theta & \cos\theta \\ \sin\theta & \tan\theta \end{pmatrix}$ is a singular matrix, where θ is any angle such that $\cos\theta \neq 0$

11 $\mathbf{A} = \frac{1}{\sqrt{2}}\begin{pmatrix} 1 & -1 \\ 1 & 1 \end{pmatrix}$ represents an anticlockwise rotation of the acute angle θ about the origin.

a Find θ.

Given that $\mathbf{B} = \begin{pmatrix} \cos 135° & -\sin 135° \\ \sin 135° & \cos 135° \end{pmatrix}$

b find the angle of rotation, about the origin, associated with the matrix **BA**.
Write down this matrix.

c Deduce that
i $\cos 135° \cos 45° - \sin 135° \sin 45° = -1$
ii $\tan 135° = -1$

12 A reflection in a line which passes through the origin maps point $A(-2, 2)$ onto itself.

a Find the matrix **P** which represents this transformation.

b State the transformation represented by the matrix $\mathbf{Q} = \begin{pmatrix} \frac{1}{2} & 0 \\ 0 & \frac{1}{2} \end{pmatrix}$

c Find the coordinates of the point B which is mapped onto point A under the combined transformation **Q** followed by **P**.

13 Given that $\mathbf{A} = \begin{pmatrix} 4 & 6 \\ 2 & 3 \end{pmatrix}$ and $\mathbf{B} = \begin{pmatrix} 3 & q \\ p & r \end{pmatrix}$ are such that $\mathbf{AB} = \mathbf{BA} = \mathbf{O}$

a find the values of p, q and r

b show that $(\mathbf{A} + \mathbf{B})^2 = \mathbf{A}^2 + \mathbf{B}^2$

FP1

14 In the diagram, $ABCD$ is a square.
The front is white (as shown), the back is black.

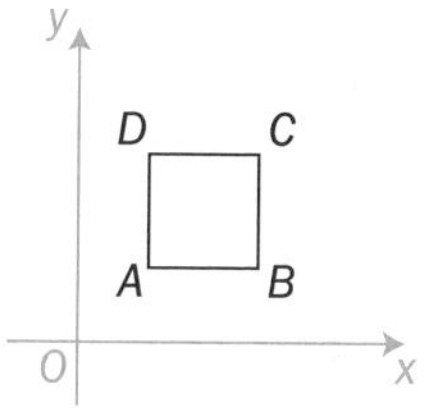

Under the matrix $\mathbf{M} = \begin{pmatrix} 4 & 3 \\ 6 & 3 \end{pmatrix}$, this square is transformed into a parallelogram with corresponding vertices A', B', C' and D'

a Given that the length $AC = 3\sqrt{2}$ units find the area of square $ABCD$.

b Hence show that parallelogram $A'B'C'D'$ has area 54 square units and state the colour of the visible face.

The matrix **N** represents a reflection in the line $y = x$.

c Find the area of the shape which is the image of the square $ABCD$ under the transformation represented by $\mathbf{NM}^{-1}$ and state the colour of its visible face.

15 **A**, **B** and **C** are 2 × 2 matrices.

a Given that $\mathbf{AB} = \mathbf{AC}$, and that **A** is non-singular, prove that $\mathbf{B} = \mathbf{C}$

b Given that $\mathbf{A} = \begin{pmatrix} 3 & 6 \\ 1 & 2 \end{pmatrix}$ and $\mathbf{B} = \begin{pmatrix} 1 & 5 \\ 0 & 1 \end{pmatrix}$, find a matrix **C**, all of whose elements are non-zero, such that $\mathbf{AB} = \mathbf{AC}$

16 Matrix $\mathbf{A} = \begin{pmatrix} 2 & 1 \\ -1 & 4 \end{pmatrix}$

a Verify that, under the transformation represented by **A**, the image P' of every point $P(k, k)$, for k any real number, also lies on the line $y = x$ and state the value of $\frac{OP'}{OP}$.

The diagram shows a square $OABC$. The image of this square is a parallelogram $OA'B'C'$, where the vertices A', B' and C' correspond to A, B and C respectively.

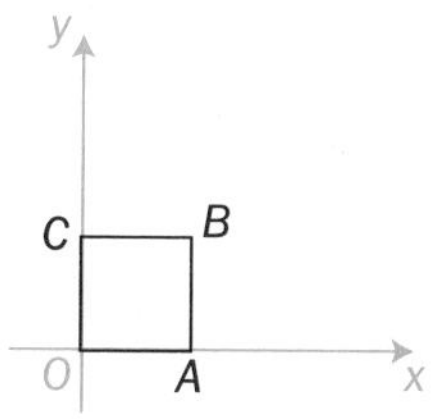

The distance $OB' = 12$ units.

b Find the length of the diagonal OB and hence show that the area of the square $OABC$ is 8 square units.

c Find the area of the parallelogram $OA'B'C'$.

FP1

5 Exit

Summary

Refer to

- The matrix $A_{r\times c}$ has r rows and c columns.
  Its order is $r \times c$ and a_{ij} is the element in the ith row and jth column. — 5.1
- The determinant of $A = \begin{pmatrix} a & b \\ c & d \end{pmatrix}$ is $ad - bc$ — 5.3

  This is written as det A or as $\left|\begin{pmatrix} a & b \\ c & d \end{pmatrix}\right|$
- A matrix is singular if its determinant is 0. — 5.4
  A is non-singular if det $A \neq 0$
- Every non-singular matrix $A = \begin{pmatrix} a & b \\ c & d \end{pmatrix}$ has an inverse matrix

  $$A^{-1} = \frac{1}{\det A}\begin{pmatrix} d & -b \\ -c & a \end{pmatrix}$$

  where $AA^{-1} = A^{-1}A = I$ and I is the identity matrix $\begin{pmatrix} 1 & 0 \\ 0 & 1 \end{pmatrix}$ — 5.4
- If A and B represent two transformations then BA represents the combined transformation A followed by B. — 5.6
- If the non-singular matrix A represents a transformation then A^{-1} represents the inverse transformation. — 5.6
- The determinant of a matrix is the area scale factor of the transformation. — 5.6
  A negative determinant indicates that the transformation represented by the matrix involves a reflection.

Links

Matrices are an indispensable tool for mathematicians to solve systems of equations and to study transformations in two and three dimensions. They also have applications in finance where they can be used to collect and store large amounts of data. The growth of internet security over recent years has led to a greater reliance on coding and transforming messages and data which can be done using matrices and their inverses.

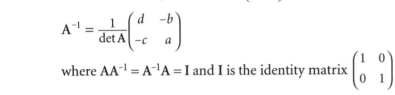

6 Series

This chapter will show you how to

- use standard results to prove summation formulae

  e.g. $\sum_{r=1}^{n} 3r(r-3) = n(n+1)(n-4)$

- efficiently evaluate series

  e.g. $\sum_{r=5}^{20} 3r(r-3)$

Before you start

You should know how to:

1 Factorise algebraic expressions completely.

e.g. Factorise $n(2n-1)(n+2) + 3n$

$$\begin{aligned} n(2n-1)(n+2) + 3n &= n[(2n-1)(n+2) + 3] \\ &= n[2n^2 + 3n + 1] \\ &= n(2n+1)(n+1) \end{aligned}$$

2 Evaluate arithmetic series.

e.g. Evaluate $\sum_{r=1}^{20} (3+2r)$

Use $S_n = \frac{n}{2}(a+l)$:

$$\begin{aligned} \sum_{r=1}^{20} (3+2r) &= 5 + 7 + 9 + \cdots + 43 \\ &= \frac{1}{2} \times 20 \times (5 + 43) \\ &= 480 \end{aligned}$$

Check in:

1 Factorise completely

a $n^3 + 4n^2 + 3n$

b $n(n+1)^2 - 4n$

See C1 for revision.

2 Evaluate

a $\sum_{r=1}^{40} (5r-1)$

b $\sum_{r=8}^{25} (3-4r)$

6.1 Summation of finite series

You can use a formula to evaluate a series.

A series is the sum of a sequence of terms.

EXAMPLE 1

a Use an appropriate formula to show that

$$\sum_{r=1}^{n} r = \frac{1}{2}n(n+1)$$

b Hence evaluate $\sum_{r=1}^{40} 4r$

Evaluate means 'find the value of'.

---

a $\sum_{r=1}^{n} r = 1 + 2 + 3 + 4 + \cdots + n$ is the sum, S_n, of the first n terms of an arithmetic series with first term $a = 1$ and common difference $d = 1$.

Hence $\sum_{r=1}^{n} r = \frac{1}{2}n[2a + (n-1)d]$

$= \frac{1}{2}n[2 \times 1 + (n-1) \times 1]$

$= \frac{1}{2}n(n+1)$, as required.

$S_n = \frac{1}{2}n[2a + (n-1)d]$ is in formula booklet (under C1).

$2 + (n-1) = n + 1$

b $\sum_{r=1}^{40} 4r = 4 \times \sum_{r=1}^{40} r$

$= 4 \times \left[\frac{1}{2} \times 40 \times (40+1)\right]$

$= 3280$

Take the constant multiplier 4 outside the summation.

Use the result from part **a** with $n = 40$.

The standard summation formulae are

$$\sum_{r=1}^{n} k = kn \quad \text{(where } k \text{ is a constant)}$$

$$\sum_{r=1}^{n} r = \frac{1}{2}n(n+1)$$

$$\sum_{r=1}^{n} r^2 = \frac{1}{6}n(n+1)(2n+1)$$

$$\sum_{r=1}^{n} r^3 = \frac{1}{4}n^2(n+1)^2, \text{ or } \left(\frac{1}{2}n(n+1)\right)^2$$

e.g. $\sum_{r=1}^{5} 2 = \underbrace{2 + 2 + 2 + 2 + 2}_{5 \text{ times}}$

$= 5 \times 2 = 10$

The result $\sum_{r=1}^{n} r = \frac{1}{2}n(n+1)$ is not explicitly stated in the formula booklet. You should learn it.

The results for $\sum r^2$ and $\sum r^3$ are in the formula booklet (under FP1).

You can use standard results to simplify expressions.

EXAMPLE 2

Use standard results to show that

$$\sum_{r=1}^{n} 3(2r^2 - 1) = n(n + 2)(2n - 1)$$

$$\sum_{r=1}^{n} 3(2r^2 - 1) = 3\sum_{r=1}^{n} (2r^2 - 1)$$

Take the constant multiplier 3 outside of the summation.

Separate into two summations: $= 3\left[2\sum_{r=1}^{n} r^2 - \sum_{r=1}^{n} 1\right]$

Use the standard results: $= 3\left[2 \times \frac{1}{6}n(n + 1)(2n + 1) - n\right]$

$\sum_{r=1}^{n} 1 = 1 + 1 + \cdots + 1 = n$

Expand the square bracket: $= n(n+1)(2n+1) - 3n$

Factorise: $= n[(n+1)(2n+1) - 3]$

$= n[2n^2 + 3n - 2]$

$= n(2n-1)(n+2)$ as required.

You can evaluate a summation which does not start with $r = 1$ by writing it as the difference of two sums.

EXAMPLE 3

Evaluate $\sum_{r=4}^{20} r^3$

The standard result for $\sum r^3$ cannot be used directly, since the given summation does not begin with $r = 1$.

Rewrite as a difference of two sums, each beginning with $r = 1$:

$$\sum_{r=4}^{20} r^3 = 4^3 + 5^3 + \cdots + 19^3 + 20^3$$

$$= (1^3 + 2^3 + 3^3 + 4^3 + 5^3 + \cdots + 19^3 + 20^3) - (1^3 + 2^3 + 3^3)$$

$$= \sum_{r=1}^{20} r^3 - \sum_{r=1}^{3} r^3$$

You add and then subtract the terms 1^3, 2^3 and 3^3.

Apply the standard result $\sum_{r=1}^{n} r^3 = \frac{1}{4}n^2(n + 1)^2$ to each summation:

$$= \frac{1}{4} \times 20^2 \times 21^2 - \frac{1}{4} \times 3^2 \times 4^2 = 44\,100 - 36 = 44\,064$$

Exercise 6.1

1 Use standard results to evaluate

a $\sum_{r=1}^{15} 2$ **b** $\sum_{r=1}^{20} 2r$ **c** $\sum_{r=1}^{10} (3r - 1)$ **d** $\sum_{r=1}^{10} 3r^2$

e $\sum_{r=1}^{15} (r^2 + 3)$ **f** $\sum_{r=1}^{40} (6r)^2$ **g** $\sum_{r=1}^{16} r^3$ **h** $\sum_{r=1}^{40} (1 - 2r^3)$

2 a Show that $(r+2)(r-2) = r^2 - 4$ for all values of r.

b Hence evaluate $\sum_{r=1}^{24} (r+2)(r-2)$

3 Evaluate each of these summations.

a $\sum_{r=1}^{12} (r-3)(r+3)$ b $\sum_{r=1}^{30} 2r(r+1)$ c $\sum_{r=1}^{18} (r+1)(r+3)$

d $\sum_{r=1}^{18} (2r+1)^2$ e $\sum_{r=1}^{20} r(r+1)^2$ f $\sum_{r=1}^{50} (4r^2-1)(2r+3)$

4 a Use standard results to show that $\sum_{r=1}^{n} (2r-1) = n^2$

b Given that the sum of the first N odd integers is greater than 800, find the smallest possible value of N.

5 Use standard results to prove that

a $\sum_{r=1}^{n} (4r+1) = n(2n+3)$ b $\sum_{r=1}^{n} (2r-3) = n(n-2)$

c $\sum_{r=1}^{n} 2r(3r-1) = 2n^2(n+1)$ d $\sum_{r=1}^{n} 3r(r+1) = n(n+1)(n+2)$

6 a Use standard results to show that

$$\sum_{r=1}^{n} (3r^2-5) = \frac{1}{2}n(n+3)(2n-3)$$

b Evaluate $\sum_{r=1}^{38} (3r^2-5)$

c Find the value of the positive integer N such that

$$\sum_{r=1}^{N} (3r^2-5) = 9N$$

7 Evaluate each series using standard results.

a $\sum_{r=6}^{15} r$ b $\sum_{r=10}^{20} (4r+1)$ c $\sum_{r=8}^{30} r^2$

d $\sum_{r=6}^{25} (2r^2-3)$ e $\sum_{r=9}^{18} r(r+2)$ f $\sum_{r=14}^{35} r(2r^2-1)$

FP1

8 Using standard results, show that

$$\sum_{r=1}^{n}(2r^3+1)=\frac{1}{2}n(n+2)(n^2+1)$$

9 **a** Use standard results to show that

$$\sum_{r=1}^{n}12r^2(r+1)=n(n+1)(n+2)(3n+1)$$

b Hence evaluate the series $\sum_{r=13}^{45}r^2(r+1)$

10 **a** Use standard results to express the sum $\sum_{r=1}^{n}r(r^2-1)$ in terms of n.

b Hence show that the sum $\sum_{r=1}^{n}4r(r^2-1)$ is the product of four consecutive integers.

11 Use standard results to express each of these summations in terms of n only. Simplify each answer as far as possible.

a $\sum_{r=1}^{2n}(r^3-r)$ **b** $\sum_{r=n+1}^{2n}6r^2$ **c** $\sum_{r=n+1}^{2n}4r^3$

12 **a** Use standard results to show that

$$\sum_{r=1}^{n}r(2r^2+k)=\frac{1}{2}n(n+1)(n^2+n+k)$$

where k is any constant.

b Hence, without the use of a calculator, show that

i $\sum_{r=1}^{175}r(2r^2+k)$ is exactly divisible by 100, where k is any integer

ii when $k=-2$, the sum $\sum_{r=1}^{1001}r(2r^2+k)$ is exactly divisible by 1000.

13 **a** Express the sum

$$T_n=n\times1^2+(n-1)\times2^2+(n-2)\times3^2+\ldots+2\times(n-1)^2+1\times n^2$$

where n is a positive integer, using sigma notation.

b Hence, by using standard results, show that

$$T_n=\frac{1}{12}n(n+1)^2(n+2)$$

FP1

Review 6

1 Use standard series to evaluate

a $\sum_{r=1}^{25}(5r-1)$

b $\sum_{r=1}^{17}(2r^2+3)$

c $\sum_{r=5}^{25}(2r)^3$

d $\sum_{r=4}^{30}(3r+1)(r-1)$

e $\sum_{r=1}^{20}(3r+1)^2$

f $\sum_{r=7}^{30}(3-2r)^2$

2 a Use standard results to show that

$$\sum_{r=1}^{n}(4r+1)(4r-1)=\frac{1}{3}n(4n+1)(4n+5)$$

for all positive integers n.

b Hence evaluate the sum $\sum_{r=9}^{28}(4r+1)(4r-1)$

3 a Use standard results to find an expression for $\sum_{r=1}^{n}r(3r-1)$

Factorise your answer as far as possible.

b Hence show that, for any integer $N > 1$,

$\sum_{r=1}^{N^2-1}r(3r-1)$ is a perfect square.

4 a Use standard results to prove that

$$\sum_{r=1}^{n}(r-3)(r-1)=\frac{1}{6}n(n-1)(2n-7)$$

b Hence evaluate $\sum_{r=17}^{29}(r-3)(r-1)$

5 a Use standard results to prove that

$$\sum_{r=1}^{n}(r+1)(r+4) = \frac{1}{3}n(n+4)(n+5)$$

b Hence show that if the positive integer N is exactly divisible by 4 then the sum $\sum_{r=1}^{N} 3(r+1)(r+4)$ is a multiple of 48.

6 a Use standard results to prove that

$$\sum_{r=1}^{n}(r-1)(2r+3) = \frac{1}{6}n(n-1)(4n+13)$$

b Hence find the value of the positive integer N for which

$$\sum_{r=1}^{N}(r-1)(2r+3) = 22N$$

7 a Use standard results to prove that

$$\sum_{r=1}^{n} 12r^2(r-1) = n(n+1)(n-1)(3n+2)$$

b Show that $r(r-1)^2 - r^2(r-1) = r - r^2$ for all r.

c Hence, or otherwise, find an expression for

$$\sum_{r=1}^{n} 12r(r-1)^2$$

factorising your answer as far as possible.

8 Prove that $\sum_{r=1}^{n} 6(r^2-1) \equiv (n-1)n(2n+5)$ [(c) Edexcel Limited 2002]

9 Prove that $\sum_{r=1}^{n}(r-1)(r+2) \equiv \frac{1}{3}(n-1)n(n+4)$ [(c) Edexcel Limited 2006]

10 a Show that $\sum_{r=1}^{n} r(r^2+1) \equiv \frac{1}{4}n(n+1)(n^2+n+2)$

b Hence evaluate $\sum_{r=7}^{25} r(r^2+1)$

FP1

6 Exit

Summary

Refer to

- $\sum_{r=1}^{n} k = kn$ (where k is a constant) — 6.1
- $\sum_{r=1}^{n} r = \frac{1}{2}n(n+1)$ — 6.1
- $\sum_{r=1}^{n} r^2 = \frac{1}{6}n(n+1)(2n+1)$ — 6.1
- $\sum_{r=1}^{n} r^3 = \frac{1}{4}n^2(n+1)^2$, or $\left(\frac{1}{2}n(n+1)\right)^2$ — 6.1

The result for $\sum_{r=1}^{n} r$ is not explicitly stated in the formula book.

You should learn it.

The formulae for $\sum r^2$ and $\sum r^3$ are in the formula booklet.

Links

Sequences and series have applications in academic subjects such as physics.

e.g. Fourier series are used in the processing of signals (such as sound waves) and are behind technologies such as musical synthesizers.

They are also used to model growth or decay of naturally occurring phenomena.

e.g. The Fibonacci sequence

1, 1, 2, 3, 5, 8, ...

defined by the iterative formula $u_{n+1} = u_n + u_{n-1}$, $u_1 = u_2 = 1$ occurs in many natural settings including growth of species and arrangements of spirals on the surface of a pine cone.

Sequences and series can be used to model long-term values of investments or to predict the length of time required to pay back a loan or a mortgage.

7 Proof

This chapter will show you how to

- prove that a mathematical statement is true for all natural numbers.

Before you start

You should know how to:

1 Factorise expressions.

e.g. Fully factorise $k(k+1)+(k+1)(k+2)$

$k(k+1)+(k+1)(k+2)=(k+1)[k+(k+2)]$
$=(k+1)(2k+2)$
$=2(k+1)^2$

2 Work with sequences and series.

e.g. For the iterative sequence $u_1=6$, $u_{n+1}=u_n+0.5$, $n\geqslant 1$, find $\sum_{r=1}^{25}u_r$

The sequence 6, 6.5, 7, 7.5, ..., is arithmetic with first term $a=6$ and common difference $d=0.5$

The sum S_n of the first n terms of an arithmetic series is $S_n=\frac{1}{2}n(2a+(n-1)d)$

$\sum_{r=1}^{25}u_r=S_{25}=\frac{1}{2}\times 25(2\times 6+(25-1)\times 0.5)$
$=300$

3 Work with matrices.

e.g. If $A=\begin{pmatrix}3 & -1\\ 1 & 2\end{pmatrix}$ find

a det A **b** A^{-1} **c** A^2

a det A $=3\times 2-1\times(-1)=7$

b $A^{-1}=\frac{1}{7}\begin{pmatrix}2 & 1\\ -1 & 3\end{pmatrix}$

c $A^2=\begin{pmatrix}3 & -1\\ 1 & 2\end{pmatrix}\begin{pmatrix}3 & -1\\ 1 & 2\end{pmatrix}$
$=\begin{pmatrix}9-1 & -3-2\\ 3+2 & -1+4\end{pmatrix}=\begin{pmatrix}8 & -5\\ 5 & 3\end{pmatrix}$

Check in:

See C1 and C2 for revision.

1 Fully factorise these expressions.

a $k^2(k+2)+k(k+2)^2$
b $(k+2)^2-k^2$
c $k(k+1)(k+6)-2k^2$
d k^3-k^2-k+1

2 **a** For the iterative sequence $u_1=2$, $u_{n+1}=\frac{1}{1+u_n}$, $n\geqslant 1$, find u_2, u_3 and u_4

Refer to Section 1.4.

b For the iterative sequence $u_1=9$, $u_{n+1}=\frac{2}{3}u_n$, $n\geqslant 1$, find the sums of these geometric series.

i $\sum_{r=1}^{10}u_r$ (to 1 d.p.) ii $\sum_{r=1}^{\infty}u_r$

3 **a** $A=\begin{pmatrix}2 & -4\\ 1 & 5\end{pmatrix}$ and $B=\begin{pmatrix}7 & 3\\ 4 & 2\end{pmatrix}$

Refer to Chapter 5.

Find i $A+B$ ii A^2 iii B^{-1}

b $A=\begin{pmatrix}1 & 0\\ 0 & 2\end{pmatrix}$

Find i A^2 ii A^3
iii Suggest an expression for A^n for any positive integer n.

c Find the value of x for which $\begin{pmatrix}9 & 4\\ 6 & x\end{pmatrix}$ is singular.

7.1 Proof by induction

A statement such as '$2^n + 3$ is a prime number' where n is a natural number may be true for some values of n and false for others.

$\mathbb{N} = \{1, 2, 3, \ldots\}$ is the set of natural numbers.

e.g. When $n = 1$, $2^n + 3 = 2^1 + 3 = 5$, is a prime number

The first few prime numbers are 2, 3, 5, 7, 11, …

So this statement is true when $n = 1$

You can express this by writing $S(1)$ is true, where $S(n)$ represents the statement '$2^n + 3$ is a prime'.

$S(2)$ is also true, since when $n = 2$

$2^n + 3 = 2^2 + 3 = 7$ is also a prime number.

You can construct a table to show the result of testing $S(n)$ for different values of n:

n	1	2	3	4	5
$2^n + 3$	5	7	11	19	35
$S(n)$	True	True	True	True	False

$35 = 5 \times 7$ is not prime.

So $S(n)$ is not true for all natural numbers n.
The first value of n for which this statement is false is $n = 5$

FP1

You can prove a statement $S(n)$ is true for *all* values of n by showing $S(n)$ satisfies two conditions.
This is known as the principle of mathematical induction.

Let $S(n)$ be any statement about the natural number n.
If (I) $S(1)$ is true
and (II) for any natural number k, $S(k + 1)$ is true whenever $S(k)$is true
Then $S(n)$ is true for *all* $n \in \mathbb{N}$

A proof by induction means showing (I) and (II) are true.

$n \in \mathbb{N}$ means n is any natural number.

Given that (I) and (II) have been shown to be true, you can see why $S(n)$ must be true for all $n \in \mathbb{N}$
By (I), $S(1)$ is true
Since $S(1)$ is true, (II) shows that $S(1 + 1)$, i.e. $S(2)$, is also true
Since $S(2)$ is true, (II) again shows that $S(2 + 1)$, or $S(3)$, is also true, and so on for all $n \in \mathbb{N}$

EXAMPLE 1

Prove by induction that $1 + 2 + 3 + \cdots + n = \frac{1}{2}n(n+1)$ for all natural numbers $n \geqslant 1$

Compare with the result in Section 6.1.

Define the statement:

Let $S(n)$ be the statement '$1 + 2 + 3 + \cdots + n = \frac{1}{2}n(n+1)$'

Show that conditions (I) and (II) are true:

(I) : When $n = 1$, $\quad 1 + 2 + 3 + \cdots + n = 1$

and $\frac{1}{2}n(n+1) = \frac{1}{2} \times 1 \times (1+1) = 1$

Hence, when $n = 1$, $\quad 1 + 2 + 3 + \cdots + n = \frac{1}{2}n(n+1)$

$\therefore$ $S(1)$ is true

When $n = 1$, the sum $1 + 2 + 3 + \ldots + n$ consists only of the first term 1.

Make an **inductive hypothesis**:

(II) : Assume that $S(k)$ is true

i.e. that $1 + 2 + 3 + \cdots + k = \frac{1}{2}k(k+1)$

where k is any natural number.

You must show that $S(k+1)$ is also true, where $S(k+1)$ states that $1 + 2 + 3 + \cdots + k + (k+1) = \frac{1}{2}(k+1)(k+2)$

Replace n with $(k+1)$ in the definition of $S(n)$.

Use algebra show that $S(k+1)$ is true:

$$1 + 2 + 3 + \cdots + k = \frac{1}{2}k(k+1)$$

$S(k)$ is assumed true. This is your 'inductive hypothesis'.

Add $(k+1)$ to both sides:

$$1 + 2 + 3 + \cdots + k + (k+1) = \frac{1}{2}k(k+1) + (k+1)$$

$$= \frac{1}{2}(k+1)(k+2)$$

Take out a factor of $\frac{1}{2}(k+1)$.

Hence $1 + 2 + 3 + \cdots + k + (k+1) = \frac{1}{2}(k+1)(k+2)$

i.e. $S(k+1)$ is true.

This shows (II) holds: $S(k+1)$ is true whenever $S(k)$ is true.

$\frac{1}{2}(k+1)(k+2)$

$= \frac{1}{2}(k+1)((k+1)+1)$

Hence, by the principle of mathematical induction, $S(n)$ is true for all $n \in \mathbb{N}$

i.e. $1 + 2 + 3 + \cdots + n = \frac{1}{2}n(n+1)$ $\quad$ for all $n \in \mathbb{N}$

FP1

EXAMPLE 2

Prove by induction that $7^n + 5$ is exactly divisible by 6 for all $n \in \mathbb{Z}^+$.

$\mathbb{Z}^+$ is the set of positive integers and is the same as $\mathbb{N}$.

Define the statement:

Let $S(n)$ be the statement '$7^n + 5$ is exactly divisible by 6'.

A number A is exactly divisible by 6 if $A = 6B$ for some integer B. i.e. A is a multiple of 6.

Show that conditions (I) and (II) are true:

(I) : When $n = 1, 7^n + 5 = 7^1 + 5$

$= 12$, clearly exactly divisible by 6

Hence $S(1)$ is true.

(II) : Assume $S(k)$ is true for any $k \in \mathbb{N}$

i.e. that $7^k + 5 = 6B$ for some integer B

Assume $7^k + 5$ is a multiple of 6.

$S(k+1)$ states that $7^{k+1} + 5$ is also a multiple of 6

Replace n with $(k + 1)$ in $S(n)$.

Use rules of indices to show that $S(k + 1)$ is true:

Express 7^{k+1} in terms of 7^k:

$$7^{k+1} + 5 = 7 \times 7^k + 5$$
$$= 7 \times (6B - 5) + 5$$

If $7^k + 5 = 6B$, $7^k = 6B - 5$

Expand the bracket and simplify:

$$= 7 \times 6B - 30$$

Avoid writing 7×6 as 42.

Take out the factor 6: $= 6(7B - 5)$

Hence $7^{k+1} + 5 = 6(7B - 5)$

$= 6C$, where $C = 7B - 5$ is an integer

Since B is an integer, so is $7B - 5$

i.e. $7^{k+1} + 5$ is a multiple of 6

Hence $S(k + 1)$ is true whenever $S(k)$ is true.

(I) and (II) are true and so by mathematical induction $S(n)$ is true for all $n \in \mathbb{N}$

i.e. $7^n + 5$ is exactly divisible by 6 for all $n \in \mathbb{Z}^+$

Exercise 7.1

You may assume throughout that n is a natural number.

1 For each statement $S(n)$, express $S(k + 1)$ in terms of the natural number k.
Simplify each statement where appropriate.

a $S(n)$: '$5^n - 1$ is a multiple of 4'

b $S(n)$: '$2^{2n-1} + 1$ is exactly divisible by 3'

c $S(n)$: '$1 \times 2 + 2 \times 5 + 3 \times 8 + \cdots + n(3n - 1) = n^2(n + 1)$'

FP1

2 Use induction to prove the following for all natural numbers n.

a $1 + 3 + 3^2 + 3^3 + \cdots + 3^n = \frac{1}{2}(3^{n+1} - 1)$

b $1^3 + 2^3 + 3^3 + \cdots + n^3 = \frac{1}{4}n^2(n+1)^2$

c $\frac{1}{1\times 2} + \frac{1}{2\times 3} + \frac{1}{3\times 4} + \cdots + \frac{1}{n(n+1)} = \frac{n}{n+1}$

3 Use induction to prove that each statement in question 1 is true for all $n \geqslant 1$.

4 Prove by induction that $\sum_{r=1}^{n} r^2 = \frac{1}{6}n(n+1)(2n+1)$ for all $n \in \mathbb{N}$.

5 A sequence is defined by the iterative formula

$$u_1 = 2$$
$$u_{n+1} = 3u_n + 2 \quad \text{for } n \geqslant 1$$

Prove by induction that $u_n = 3^n - 1$ for all $n \geqslant 1$

6 $\mathbf{A} = \begin{pmatrix} 3 & 4 \\ -1 & -1 \end{pmatrix}$

a Prove by induction that $\mathbf{A}^n = \begin{pmatrix} 2n+1 & 4n \\ -n & 1-2n \end{pmatrix}$ for all $n \in \mathbb{N}$.

b Use the result of part **a** to show that $\det(\mathbf{A}^n) = 1$ for all positive integers n.

7 Use induction to prove that $6^n + 4$ is exactly divisible by 10 for all $n \in \mathbb{N}$.

8 Given that $x_n = 3^{2n} - 1$ and $y_n = 3^{2n-1} + 1$

a use induction to prove that, for all $n \in \mathbb{N}$,

i x_n is a multiple of 8 **ii** y_n is exactly divisible by 4

b by simplifying the expression $x_n + 2y_n$, or otherwise, prove that $5 \times 3^{2n-1} + 1$ is a multiple of 8 for all $n \geqslant 1$.

9 Let $T_n = \frac{1}{2} + \frac{2}{3} + \cdots + \frac{n}{n+1}$

a Use induction to prove that $T_n < n$ for all $n \geqslant 1$.

b Show that $(k+1)^3 > k^3 + 3k^2 + 2k + 1$ for all natural numbers k.

Hence prove by induction that $T_n \leqslant \frac{n^2}{n+1}$ for all $n \geqslant 1$.

FP1

Review 7

1 Use induction to prove

a $1 \times 8 + 2 \times 11 + 3 \times 14 + \cdots + n(3n+5) = n(n+1)(n+3)$ for all $n \geqslant 1$

b $\sum_{r=1}^{n}(2r+1)(6r-1) = n(2n+1)(2n+3)$ for all $n \in \mathbb{N}$

2 Prove by induction that

$$1 \times 2 \times 3 + 2 \times 3 \times 4 + \cdots + n(n+1)(n+2) = \frac{1}{4}n(n+1)(n+2)(n+3) \text{ for all } n \in \mathbb{N}.$$

3 a Use induction to prove that

i $4^n + 2$ is a multiple of 6 for all $n \in \mathbb{N}$

ii $5 \times 9^n + 3$ is exactly divisible by 12 for all $n \geqslant 1$.

b Deduce that $5(9^n + 1) + 4^n$ is exactly divisible by 6 for all $n \geqslant 1$.

4 $\mathbf{A} = \begin{pmatrix} 2 & 0 \\ 3 & 1 \end{pmatrix}$

a Prove by induction that $\mathbf{A}^n = \begin{pmatrix} 2^n & 0 \\ 3 \times (2^n - 1) & 1 \end{pmatrix}$ for all $n \geqslant 1$.

b Hence find the value of the positive integer N such the sum of all the elements of $\mathbf{A}^N$ is 2046.

5 A sequence is defined by the iterative formula

$u_1 = 1$ and $u_{n+1} = \frac{1}{5}(2u_n - 9)$ for $n \geqslant 1$

a Prove by induction that $u_n = 10\left(\frac{2}{5}\right)^n - 3$ for all $n \in \mathbb{N}$.

b Hence find the value of $\sum_{n=1}^{\infty}(u_n + 3)$

6 a Find constants A, B and C such that
$2k^3 + 9k^2 + 3k - 4 \equiv (k+1)(Ak^2 + Bk + C)$

b Prove by induction that $\sum_{r=1}^{n}(3r^2 - 5) = \frac{1}{2}n(n+3)(2n-3)$ for all $n \geqslant 1$

7 A sequence is defined by the iterative formula

$$u_{n+1} = u_n + \frac{5}{2}, \quad n \geqslant 1$$

$$u_1 = \frac{3}{2}$$

a Prove by induction that $u_n = \frac{5}{2}n - 1$ for all $n \geqslant 1$.

b Hence calculate the sum of the first 20 terms of this sequence.

8 The matrix $\mathbf{A} = \begin{pmatrix} 2 & 0 \\ 0 & 3 \end{pmatrix}$

a Show that $\mathbf{A}^2 = \begin{pmatrix} 2^2 & 0 \\ 0 & 3^2 \end{pmatrix}$

b Prove by induction that $\mathbf{A}^n = \begin{pmatrix} 2^n & 0 \\ 0 & 3^n \end{pmatrix}$ for all $n \in \mathbb{N}$.

c Hence find an expression for $(\mathbf{A}^n)^{-1}$.

Give your answer in the form $\begin{pmatrix} p^n & c \\ b & q^n \end{pmatrix}$ for rational numbers b, c, p and q to be stated.

9 Use induction to prove that $2^{3n-1} + 3$ is a multiple of 7 for all $n \in \mathbb{N}$.

10 A sequence is defined by the iterative formula

$$u_1 = 5, \qquad u_{n+1} = 1.2u_n + 0.2, \quad n \geqslant 1$$

a Prove by induction that the general term of this sequence is given by $u_n = 5 \times 1.2^n - 1$

b Hence calculate $\sum_{r=1}^{15} u_r$. Give your answer to three significant figures.

11 $\mathbf{A} = \begin{pmatrix} -1 & 2 \\ -2 & 3 \end{pmatrix}$

a Prove by induction that $\mathbf{A}^n = \begin{pmatrix} 1-2n & 2n \\ -2n & 2n+1 \end{pmatrix}$ for all $n \in \mathbb{N}$.

b Hence find an expression for $(\mathbf{A}^n)^{-1}$.

FP1

12 Use induction to prove each of the following for all $n \geqslant 1$.

a $1 + 3 + 5 + \cdots + (2n - 1) = n^2$

b $1 \times 2 + 2 \times 3 + 3 \times 4 + \cdots + n(n+1) = \frac{1}{3}n(n+1)(n+2)$

c $\sum_{r=1}^{n} r(r+2) = \frac{1}{6}n(n+1)(2n+7)$

d $\sum_{r=1}^{n} \frac{1}{(r+1)(r+2)} = \frac{n}{2(n+2)}$

13 The sequence u_n is defined by the formula $u_n = 3^{2n} - 4^n$

a Show that $u_{k+1} - 4u_k = 5 \times 9^k$, where k is any natural number.

b Hence prove by induction that $3^{2n} - 4^n$ is a multiple of 5 for all $n \geqslant 1$.

14 a Prove by induction that if **P** and **Q** are any pair of 2×2 matrices such that $\mathbf{PQ} = \mathbf{I}$, where **I** is the identity matrix, then $\mathbf{P}^n\mathbf{Q}^n = \mathbf{I}$ for all $n \in \mathbb{N}$.

b Deduce that $(\mathbf{P}^n)^{-1} = (\mathbf{P}^{-1})^n$ for any $n \in \mathbb{N}$, where **P** is any non-singular 2×2 matrix.

c Given that $\mathbf{P} = \begin{pmatrix} 7 & -6 \\ 6 & -5 \end{pmatrix}$

i prove by induction that $\mathbf{P}^n = \begin{pmatrix} 6n+1 & -6n \\ 6n & -6n+1 \end{pmatrix}$ for all $n \in \mathbb{N}$

ii hence find an expression for $(\mathbf{P}^{-1})^n$ for any $n \in \mathbb{N}$.

15 Prove by induction that
$1 \times 3 + 2 \times 4 + 4 \times 5 + \cdots + 2^{n-1}(n+2) = 2^n(n+1) - 1$ for all $n \in \mathbb{N}$.

16 Given that $u_n = 5^n + 9^n + 2$

a express $u_{n+1} - 5u_n$, simplifying your answer as far as possible

b hence prove by induction that $5^n + 9^n + 2$ is exactly divisible by 4 for all $n \geqslant 1$.

17 The matrix $\mathrm{A} = \begin{pmatrix} 2 & 0 \\ -1 & 1 \end{pmatrix}$

a Find A^2.

b Prove by induction that $\mathrm{A}^n = \begin{pmatrix} 2^n & 0 \\ 1-2^n & 1 \end{pmatrix}$ for all $n \in \mathbb{N}$.

c Given that $\mathrm{A}^N = \begin{pmatrix} a & 0 \\ -1023 & 1 \end{pmatrix}$, where N is a natural number,

find the value of a and the value of N.

18 A sequence is defined by the iterative formula

$$u_1 = \frac{1}{2}, \quad u_{n+1} = \frac{n}{n+u_n}, \quad n \geqslant 1$$

Prove by induction that $u_n = \dfrac{n}{n+1}$ for all $n \in \mathbb{N}$.

19 a Use induction to prove that

$$\frac{1}{1\times 4} + \frac{1}{4\times 7} + \frac{1}{7\times 10} + \cdots + \frac{1}{(3n-2)(3n+1)} = \frac{n}{3n+1} \quad \text{for all } n \geqslant 1$$

b Hence evaluate the sum $\dfrac{1}{1\times 4} + \dfrac{1}{4\times 7} + \dfrac{1}{7\times 10} + \cdots + \dfrac{1}{46\times 49}$,
giving your answer in the form q^2 where $q > 0$ is a rational number to be stated.

20 Use the result det (**AB**) = det **A** × det **B** for any pair of 2 × 2 matrices to prove by induction that det ($\mathbf{A}^n$) = (det **A**)^n for all $n \geqslant 1$.

21 a Prove directly that $2(3^k - 1)$ is a multiple of 4 for any $k \in \mathbb{N}$

b Hence prove by induction that $(2n-1)3^n + 1$ is exactly divisible by 4 for all $n \in \mathbb{N}$

22 A sequence is defined by the iterative formula

$$u_1 = 3, \quad u_{n+1} = \frac{nu_n + 2}{n+1}, \quad n \geqslant 1$$

a Prove by induction that

i $u_n > 2$ for all $n \in \mathbb{N}$

ii $u_n < 4$ for all $n \in \mathbb{N}$

It is given that $u_{n+1} < u_n$ for all $n \geqslant 1$

b State the value of N for which u_N is an integer.

23 $\mathrm{f}(n) = (2n+1)7^n - 1$
Prove by induction that, for all positive integers n, $\mathrm{f}(n)$ is divisible by 4

[(c) Edexcel Limited 2003]

24 Given that $u_n = 4 \times 2^n + 3 \times 9^n$

a show that $u_{n+1} - 2u_n = 21 \times 9^n$ and hence prove by induction that u_n is a multiple of 7 for all $n \geqslant 1$

b prove by induction that for all $n \geqslant 1$, u_n is *not* exactly divisible by 3.

7 Exit

Summary

Refer to

- Given a statement $S(n)$ about the natural number n — 7.1
  - If $S(1)$ is true
  - and for any natural number k, $S(k+1)$ is true whenever $S(k)$ is true
  - then $S(n)$ is true for *all* $n \in \mathbb{N}$.
- This is known as the principle of mathematical induction. — 7.1

Links

Our everyday lives depend on an inductive type of reasoning.

e.g. The sun rose yesterday and the day before that and has done so for millions of years. Based on this experience, we therefore strongly believe that the sun will rise tomorrow.

Mathematical induction is a formalisation of this idea. It is a highly effective technique when trying to prove that a statement is true for all natural numbers. Although not every such statement can be proved using this technique, induction was essentially the tool by which Professor Andrew Wiles demonstrated, in 1996, the truth of *Fermat's Last Theorem*, a problem which had tormented mathematicians for over 350 years.

FP1

1 a Completely factorise $4(k+1)^2 - 1$

b Hence prove by induction that

$$\sum_{r=1}^{n} \frac{1}{4r^2-1} = \frac{n}{2n+1} \text{ for all } n \in \mathbb{N}$$

2 a Using standard results, or otherwise, show that

$$\sum_{r=1}^{n} (2r-1)^2 = \frac{1}{3}n(2n+1)(2n-1)$$

b Hence calculate the sum $39^2 + 41^2 + 43^2 + \cdots + 59^2$

3 Matrices $\mathbf{P} = \begin{pmatrix} -2 & 3 \\ 6 & 1 \end{pmatrix}$ and $\mathbf{Q} = \begin{pmatrix} x & 2 \\ 4 & x^2 \end{pmatrix}$, where x is a real number, are such that $\mathbf{PQ} = \mathbf{QP}$. Given that $\mathbf{Q}$ is non-singular, find the value of x.

4 Prove by induction that $5^{2n-1} + 3$ is exactly divisible by 8 for all positive integers n.

5 $\mathbf{A} = \begin{pmatrix} -2 & -4.5 \\ 2 & 4 \end{pmatrix}$

a Prove by induction that $\mathbf{A}^n = \begin{pmatrix} 1-3n & -4.5n \\ 2n & 3n+1 \end{pmatrix}$ for all $n \geqslant 1$.

b Hence show that, for any $n \geqslant 1$, the matrix $\mathbf{A}^n - \mathbf{I}$ is singular, where $\mathbf{I} = \begin{pmatrix} 1 & 0 \\ 0 & 1 \end{pmatrix}$

6 a Show that $\sum_{r=1}^{n} (r+1)(r+5) = \frac{1}{6}n(n+7)(2n+7)$

b Hence calculate the value of $\sum_{r=10}^{40} (r+1)(r+5)$ [(c) Edexcel Limited 2004]

7 $\mathbf{P} = \begin{pmatrix} 3 & x \\ 1 & 5 \end{pmatrix}$ where x is a constant.

a Find an expression for $\mathbf{P}^2$.

b Given that $\mathbf{P}^2 \begin{pmatrix} -2 \\ 1 \end{pmatrix} = \begin{pmatrix} -6 \\ k \end{pmatrix}$, where k is a constant,

show that $x = 2$ and find the value of k.

c Hence, or otherwise, find $\mathbf{P}^3 \begin{pmatrix} -2 \\ 1 \end{pmatrix}$

8 a Use standard results to show that $\sum_{r=1}^{n} r(3r+1) = n(n+1)^2$

b Deduce that $10 \times 31 + 11 \times 34 + \cdots + 49 \times 148 = 320 \times 380$, showing your working.

9 a Prove by induction that $\sum_{r=1}^{n} \frac{2r+1}{r^2(r+1)^2} = 1 - \frac{1}{(n+1)^2}$ for all $n \geqslant 1$

b Evaluate $\sum_{r=12}^{23} \frac{2r+1}{r^2(r+1)^2}$, giving your answer as a fraction in its simplest terms.

10 The matrix **A** represents a reflection in the x-axis.
The matrix **B** represents a reflection in the line $y = x$

a Write down the matrices **A** and **B**.

b Using matrix multiplication, find the matrix **C** which represents the combined transformation of **A** followed by **B**.

c State a single transformation represented by

i **C** ii **CBA**

11 A sequence is defined by the iterative formula

$u_1 = 5,\ u_{n+1} = u_n + 6(2n+1),\ n \geqslant 1$

a Prove by induction that the general term is given by $u_n = 6n^2 - 1$

b Hence, using standard results, find an expression for $\sum_{r=1}^{n} u_r$ in terms of n. Factorise your answer as far as possible.

12 a Prove by induction that $4^n - 1$ is a multiple of 3 for all $n \geqslant 1$.

b Hence prove by induction that $(3n-1) \times 4^n + 1$ is exactly divisible by 9 for all positive integers n.

13 a Simplify the expression $\mathbf{A}(\mathbf{A}\mathbf{B}^{-1}\mathbf{A})^{-1}\mathbf{A}$, where **A** and **B** are any 2×2 non-singular matrices.

A 2×2 matrix **A** is self-inverse if $\mathbf{A}^{-1} = \mathbf{A}$

b Prove that if **A** is self-inverse then so is the matrix $\mathbf{B}\mathbf{A}\mathbf{B}^{-1}$, where **B** is any non-singular 2×2 matrix.

14 Using standard results

a show that $\sum_{r=1}^{n} 2r(2r^2+1) = n(n+1)(1+n+n^2)$

b find an expression for $\sum_{r=1}^{n} 2r^2(1-r)$

Factorise your answer as far as possible.

15 $\mathbf{A} = \begin{pmatrix} 3 & 2 \\ 1 & 4 \end{pmatrix}$

a Show that $\mathbf{A}\begin{pmatrix} x \\ x \end{pmatrix} = 5\begin{pmatrix} x \\ x \end{pmatrix}$ for any real number x.

b i Verify that $\mathbf{A}^2 - 7\mathbf{A} + 10\mathbf{I} = \mathbf{O}$, where $\mathbf{I} = \begin{pmatrix} 1 & 0 \\ 0 & 1 \end{pmatrix}$ and $\mathbf{O} = \begin{pmatrix} 0 & 0 \\ 0 & 0 \end{pmatrix}$

ii Hence, or otherwise, express $\mathbf{A}^3$ in the form $p\mathbf{A} + q\mathbf{I}$ for constants p and q to be stated.

c Find $\mathbf{A}^3\begin{pmatrix} 0.4 \\ 0.4 \end{pmatrix}$

16 **a** Express $\frac{6x+10}{x+3}$ in the form $p + \frac{q}{x+3}$ where p and q are integers to be found.

The sequence of real numbers $u_1, u_2, u_3, \cdots$ is such that $u_1 = 5.2$ and $u_{n+1} = \frac{6u_n + 10}{u_n + 3}$

b Prove by induction that $u_n > 5$ for all $n \in \mathbb{N}$.

[(c) Edexcel Limited 2005]

17 In the diagram, the square $OABC$, where B has coordinates (k, k) for $k > 0$, a constant, contains the parallelogram $OA'BC'$

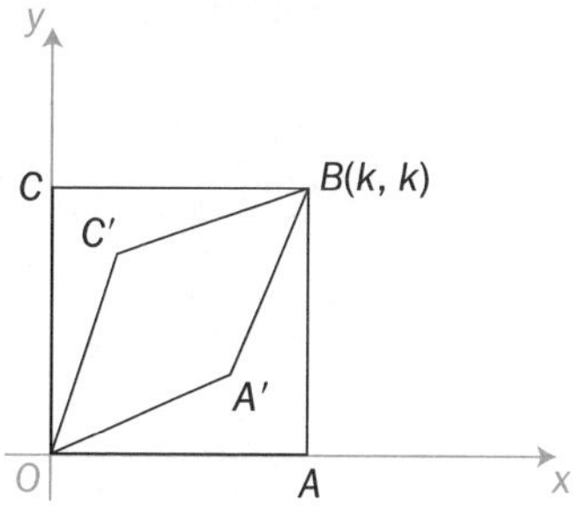

Under matrix $\mathbf{M} = \begin{pmatrix} 0.8 & p \\ 0.3 & q \end{pmatrix}$, where p and q are constants, points A and C are mapped to points A' and C' respectively, and point B is mapped to itself.

a Show that $p = 0.2$ and $q = 0.7$

b Calculate det $\mathbf{M}$ and hence

i find, in terms of k, the area of the quadrilateral $OABA'$,

ii explain why the shape whose image is the square $OABC$ under $\mathbf{M}$ cannot be completely enclosed by this square.

18 $\mathbf{P} = \begin{pmatrix} -3 & 2 \\ -8 & 5 \end{pmatrix}$

a Prove by induction that $\mathbf{P}^n = \begin{pmatrix} 1-4n & 2n \\ -8n & 4n+1 \end{pmatrix}$ for all $n \geqslant 1$.

b Hence express $\mathbf{P}^{2n} - \mathbf{P}^n$ in the form $2n\mathbf{A}$ for $\mathbf{A}$, a matrix to be found.

c Show that $\mathbf{A}^2 = \mathbf{O}$ and hence deduce that $\frac{1}{2}(\mathbf{P}^{4n} + \mathbf{P}^{2n}) = \mathbf{P}^{3n}$

19

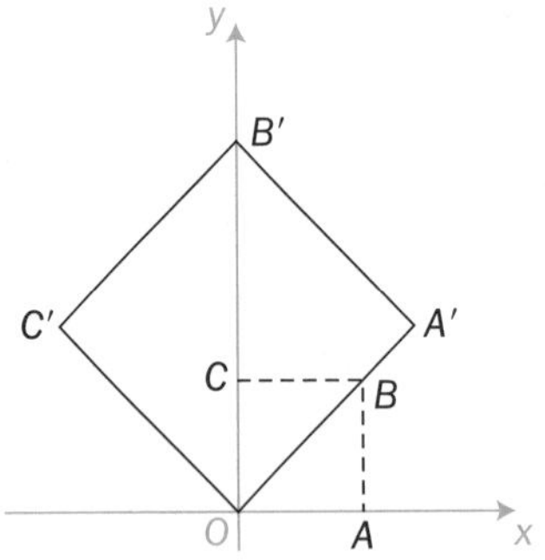

In the diagram, the unit square $OABC$ is transformed into the square $OA'B'C'$ by an enlargement scale factor $3\sqrt{2}$ followed by an anticlockwise rotation of 45°, about the origin O.

a Show that the matrix $\mathbf{M}$ which represents this combined transformation is given by $\mathbf{M} = \begin{pmatrix} 3 & -3 \\ 3 & 3 \end{pmatrix}$

b Find $\mathbf{M}^{-1}$ and describe, in the correct order, a sequence of transformations this matrix represents.

c The matrix $\mathbf{M}^{-1}$ is applied to the unit square $OABC$.

i Show, on a single sketch, the unit square $OABC$ and its image under $\mathbf{M}^{-1}$.
ii Find the area of the region common to both shapes.

20 a Use induction to prove that $u_n = 2^{2n-1} + 1$ is a multiple of 3 for all $n \geqslant 1$.

b i Given that $f(n) = 3^n + 4^n - 1$ express $f(n+1) - 3f(n)$ in terms of u_n.
ii Hence prove by induction that $f(n)$ is exactly divisible by 6 for all $n \in \mathbb{N}$.

Answers

Chapter 1

Exercise 1.1

1 a $x = -7, x = -3$ b $x = 1, x = 13$

c $x = \frac{3}{2}, x = -5$ d $x = -2, x = 11$

2 a $x = -4 \pm \sqrt{19}$ b $x = \frac{5}{2} \pm \frac{1}{2}\sqrt{17}$

c $x = -\frac{5}{4} \pm \frac{1}{4}\sqrt{33}$

3 a $x = 2 \pm \sqrt{11}$ b $x = \frac{-3 \pm \sqrt{33}}{4}$

c $x = -\frac{1}{3}, x = -2$

4 a $x = 4, x = \frac{3}{2}$ b $x = \frac{-2 \pm 2\sqrt{3}}{3}$

c $x = \frac{2 \pm \sqrt{10}}{3}$

5 a $x = 2, y = 1$, or $x = \frac{4}{3}, y = \frac{7}{3}$

b $x = 12 \pm \sqrt{46}, y = 4 \pm \sqrt{46}$

c $x = 1 + \sqrt{2}, y = 1 - \sqrt{2}$ or $x = 1 - \sqrt{2}, y = 1 + \sqrt{2}$

d $x = 1, y = 1$ or $x = \frac{4}{9}, y = -\frac{2}{3}$

6 a $x = 3, -1, 4$ b $x = -2, 1, -\frac{3}{2}$

c $x = 2, -4$ d $x = -1$

Exercise 1.2

1 a $\frac{1+\sqrt{3}}{2}$ b 1

c $\frac{2}{3}\sqrt{3}$ d $\frac{1}{4}$

2 3

3 a Angle $C = \frac{1}{4}\pi$, Angle $B = \frac{1}{12}\pi$

b Area $= \frac{1}{2}(3 - \sqrt{3})$

Exercise 1.3

1 a

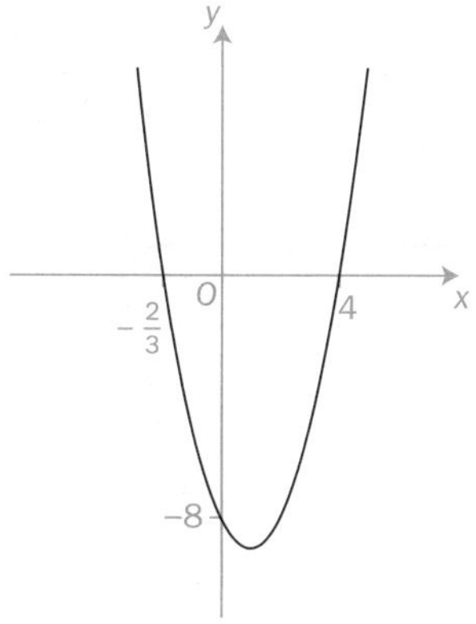

b

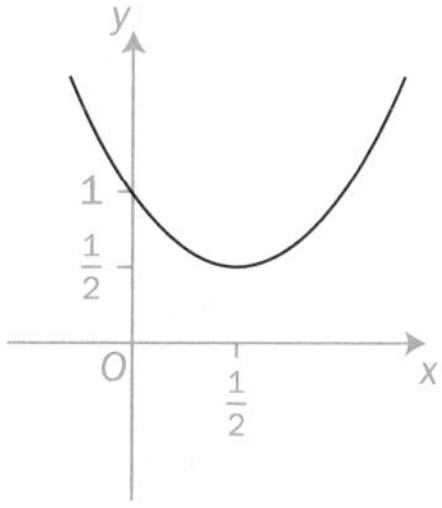

c

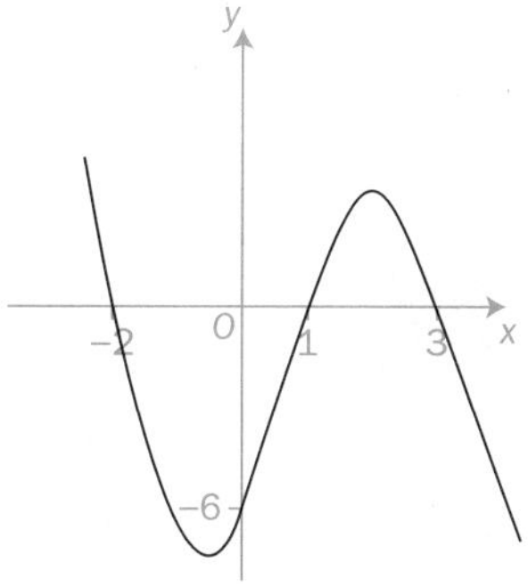

2 a

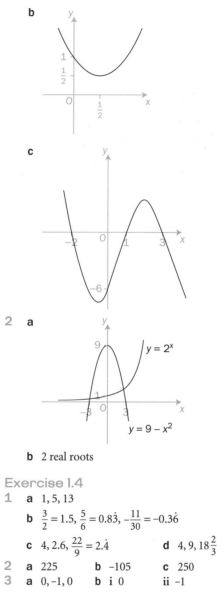

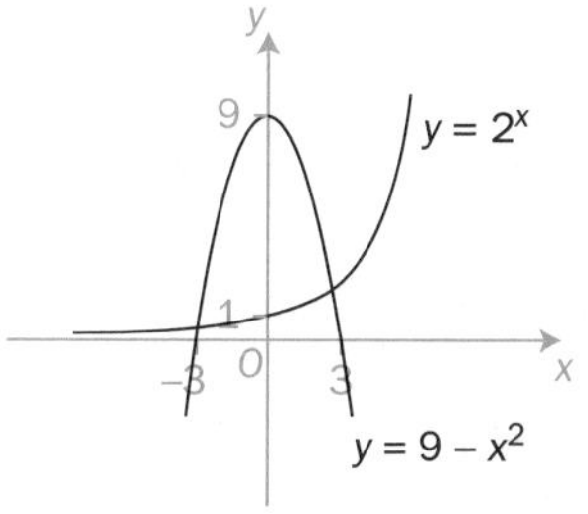

b 2 real roots

Exercise 1.4

1 a 1, 5, 13

b $\frac{3}{2} = 1.5, \frac{5}{6} = 0.8\dot{3}, -\frac{11}{30} = -0.3\dot{6}$

c $4, 2.6, \frac{22}{9} = 2.\dot{4}$ d $4, 9, 18\frac{2}{3}$

2 a 225 b −105 c 250 d 300

3 a 0, −1, 0 b i 0 ii −1 iii 0

Exercise 1.5

1 a Tangent: $y = 3x - 2$
Normal: $3y + x = 14$

b Tangent: $y = 2x + 3$
Normal: $y = -\frac{1}{2}x + \frac{1}{2}$

c Tangent: $y = 3x - 3$
Normal: $y = -\frac{1}{3}x + \frac{1}{3}$

d Tangent: $y = \frac{13}{2}x - 23$
Normal: $y = -\frac{2}{13}x + \frac{47}{13}$

FP1

e Tangent: $y=\frac{1}{3}x+\frac{8}{3}$
Normal: $y=-3x-24$
f Tangent: $y=7x-4$
Normal: $7y+x=22$

2 b -5

3 a $f'(x)=\sqrt{3}x^{-\frac{1}{2}}$
b $f'(x)=-4x^{-5}+2x^{-3}$
c $f'(x)=2x+1-x^{-2}$

4 b $13y+x+29=0$
c $x=-2, x=\frac{4}{3}$

5 a $y=7x-5$ b $\left(\frac{1}{2},-\frac{1}{4}\right),\left(-\frac{1}{2},-\frac{7}{4}\right)$
c $x=0$
d

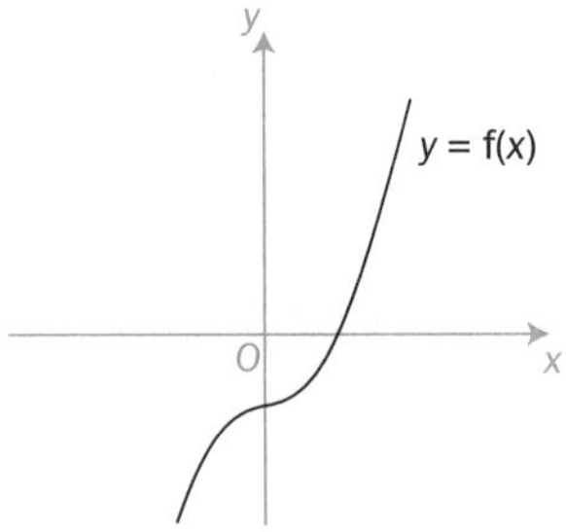

Chapter 2

Before you start

1 a $x=3, x=-4$ b $x=0, x=\frac{2}{3}$ c $x=\frac{1}{2}$

2 a $x=0, y=3$ or $x=2, y=1$
b $x=\frac{1}{2}, y=\frac{1}{2}$ or $x=-\frac{1}{2}, y=-\frac{1}{2}$
c $x=2, y=\pm1$ or $x=-2, y=\pm1$

3 $x=-1,\ x=\frac{2}{3}, x=2$

Exercise 2.1

1 a $\pm4i$ b $\pm6i$ c $\pm10i$
d $\pm\sqrt{20}i$

2 a $x=\pm2i$ b $x=\pm7i$ c $x=\pm15i$
d $x=\pm5i$ e $\pm3i$ f $x=\pm\frac{1}{2}i$
g $x=\pm\frac{2}{3}i$ h $x=\pm\frac{\sqrt{2}}{3}i$

3 a $x=\pm2\sqrt{2}i$ b $x=\pm3\sqrt{2}i$ c $\pm2\sqrt{3}i$
d $x=\pm3\sqrt{5}i$ e $x=\pm2\sqrt{6}i$ f $\pm2\sqrt{7}i$
g $x=\pm\frac{\sqrt{3}}{6}i$ h $x=\pm\frac{\sqrt{3}}{2}i$

4 a $x=\pm ai$ b $x=\pm3ai$
c $x=\pm a^2i$ d $x=\pm\frac{1}{a}i$

5 b $a=2, b=25$

6 a $x=\pm\sqrt{\sqrt{2}i}$ b $x=\pm2, x=\pm2i$
c $x=\pm\frac{1}{3}i$ d $x=\pm5i$

Exercise 2.2

1 a $\text{Re}(z)=3, \text{Im}(z)=2$
b $\text{Re}(z)=4, \text{Im}(z)=-5$
c $\text{Re}(z)=-1, \text{Im}(z)=4$
d $\text{Re}(z)=2, \text{Im}(z)=-6$
e $\text{Re}(z)=\frac{1}{2}, \text{Im}(z)=\frac{1}{3}$
f $\text{Re}(z)=-5, \text{Im}(z)=-3$
g $\text{Re}(z)=-7, \text{Im}(z)=0$
h $\text{Re}(z)=\sqrt{2}, \text{Im}(z)=0$
i $\text{Re}(z)=0, \text{Im}(z)=\pm2$

2 a $\text{Re}(z)=2, \text{Im}(z)=-3$
b $\text{Re}(z)=-4, \text{Im}(z)=-2$
c $\text{Re}(z)=0, \text{Im}(z)=\frac{1}{3}$
d $\text{Re}(z)=\frac{1}{3}, \text{Im}(z)=\frac{2}{3}$
e $\text{Re}(z)=\sqrt{2}, \text{Im}(z)=-2$
f $\text{Re}(z)=\sqrt{3}, \text{Im}(z)=\sqrt{2}$

3 a $z=3+9i, z=-3+9i$
b $z=0, z=4+4i$

4 a 2 complex roots b 2 distinct real roots
c 2 distinct real roots d 2 equal real roots
e 2 complex roots

5 a $z=-2\pm i$ b $z=1\pm2i$
c $z=2\pm3i$ d $z=-\frac{1}{2}\pm\frac{3}{2}i$
e $z=\frac{1}{5}\pm\frac{3}{5}i$ f $z=\frac{3}{4}\pm\frac{5}{4}i$

6 a $z=-\frac{3}{2}\pm\frac{\sqrt{3}}{2}i$ b $z=\frac{1}{4}\pm\frac{\sqrt{7}}{4}i$
c $z=-2\pm\sqrt{2}i$ d $z=\frac{5}{8}\pm\frac{\sqrt{23}}{8}i$
e $z=\frac{5}{6}\pm\frac{\sqrt{11}}{6}i$ f $z=\frac{11}{10}\pm\frac{\sqrt{19}}{10}i$

7 a $z=\frac{1}{2}\pm\frac{\sqrt{3}}{2}i$ b $\frac{1}{3}\pm\frac{2\sqrt{2}}{3}i$
c $z=-1\pm\frac{\sqrt{2}}{2}i$ d $z=-\frac{3}{2}\pm\frac{3\sqrt{3}}{2}i$
e $-1\pm\frac{\sqrt{6}}{2}i$ f $z=\frac{1}{3}\pm\frac{\sqrt{3}}{3}i$

8 a $1\pm\sqrt{3}i$ b $-3\pm\sqrt{3}i$
c $3\pm2\sqrt{3}i$ d $-1\pm4\sqrt{3}i$

9 a $z=\frac{3}{2}\pm\frac{\sqrt{11}}{2}i$ b $z=\frac{1}{2}\pm\frac{\sqrt{13}}{2}i$
c $z=\frac{1}{3}\pm\frac{\sqrt{14}}{3}i$ d $z=\frac{3}{2}\pm\frac{\sqrt{7}}{2}i$
e $z=\frac{3}{2}\pm\frac{\sqrt{3}}{2}i$ f $z=1\pm\frac{\sqrt{10}}{2}i$

10 b $z=1, z=-\frac{1}{2}\pm\frac{\sqrt{3}}{2}i$

Exercise 2.3

1 **a** $7+5i$ **b** $4-3i$ **c** $-6-5i$
d $1+0i$ **e** $6+9i$ **f** $-12+3i$
g $1+5i$ **h** $2-4i$ **j** $-6+5i$

2 **a** $z=2+i$ **b** $z=1+\frac{3}{2}i$ **c** $z=0+\frac{1}{2}i$

3 **a** $z=-5+3i,\ w=6+i$
b $z=1+i,\ w=0+3i$
c $z=2-i,\ w=-1+2i$

4 **a** $-14+8i$ **b** $2+26i$ **c** $3-4i$

5 **a** $\frac{13}{5}+\frac{1}{5}i$ **b** $\frac{1}{2}+\frac{1}{2}i$ **c** $-2-\frac{3}{2}i$

6 **a** $z=2+3i$ **b** $z=4-i$ **c** $z=\frac{1}{2}+\frac{3}{2}i$
d $z=1-i$ **e** $z=-\frac{3}{10}-\frac{9}{10}i$ **f** $z=\frac{3}{5}+\frac{4}{5}i$

7 **a** $z=4+i,\ w=-2+3i$
b $z=-4+4i,\ w=3-4i$
c $z=4-2i,\ w=4-i$

8 $z=\pm i$

9 **a** $z=3+2i,\ w=2+3i$
b $z=1+2i,\ w=3-i$
c $z=2+3i,\ w=1-2i$ or $z=2+i,\ w=3-2i$

Exercise 2.4

1 **a** $a=7, b=2$ **b** $a=4, b=-2$
c $a=3, b=-8$ **d** $a=3, b=5$
e $a=2, b=4$ **f** $a=-2, b=0$

2 **a** $a=3, b=-2$ **b** $a=6, b=\frac{1}{2}$
c $a=0, b=0$ or $a=4, b=-4$
d $a=\pm\sqrt{3}, b=0$
e $a=1, b=-2$ or $a=-1, b=2$
f $a=2, b=\pm 4$

3 **b** $q=10$

4 **a** $z=\pm(4+i)$ **b** $z=\pm(2+5i)$
c $z=\pm(4-2i)$ **d** $z=\pm(2+2i)$
e $z=\pm\left(\frac{3}{2}-\frac{1}{2}i\right)$ **f** $z=\pm\left(\frac{1}{2}+2i\right)$

5 **a** $3-7i$ **b** $-2-5i$ **c** $3+9i$
d $-1+i$ **e** $-4-5i$ **f** $-\frac{3}{4}i$

6 **a** $4-8i$ **b** $8i$ **c** $10-4i$
d $4+2i$ **e** $-6+6i$

7 **a** $z=1+3i$ **b** $z=3-2i$ **c** $5-2i$
d $z=4-i$ **e** $3+3i$ **f** $z=\frac{2}{5}+\frac{3}{5}i$

8 **a** $z=6+4i,\ w=2-3i$
b $z=4-i,\ w=5+3i$

11 **a** i

12 **a** $b=\frac{1}{2}(-c+\sqrt{c^2+d^2})$
c $13^4=(\pm 119)^2+(\pm 120)^2$

Exercise 2.5

1 **a** $p=2+3i,\ u=3-i,\ v=-2+i,\ w=3-2i,\ z=-3-2i$
b w

2

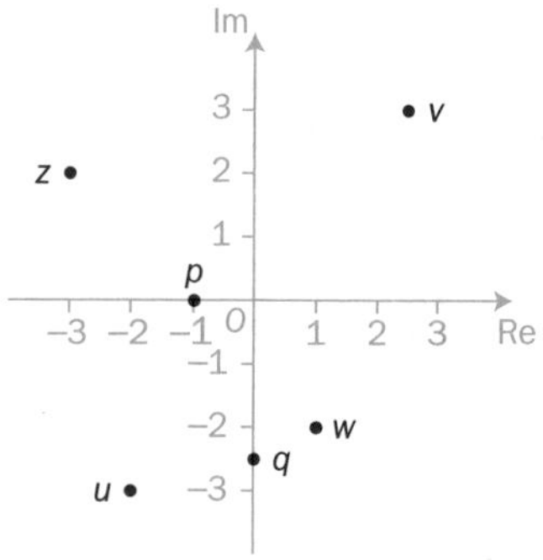

3 **a** i, ii

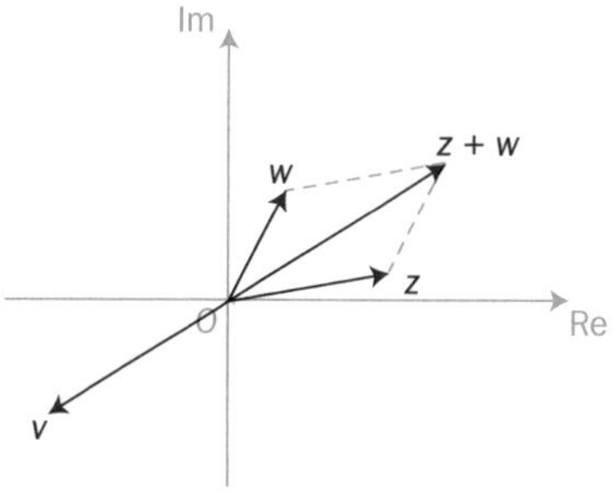

b 3rd quadrant

4 **a** i–iii **b** Trapezium

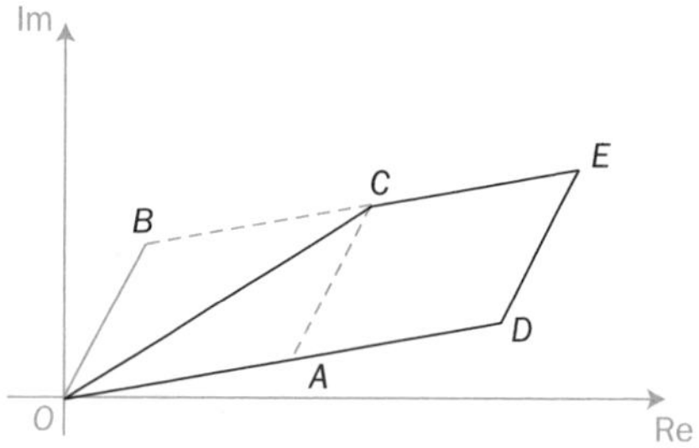

6 **a** i – ii **b** Rhombus

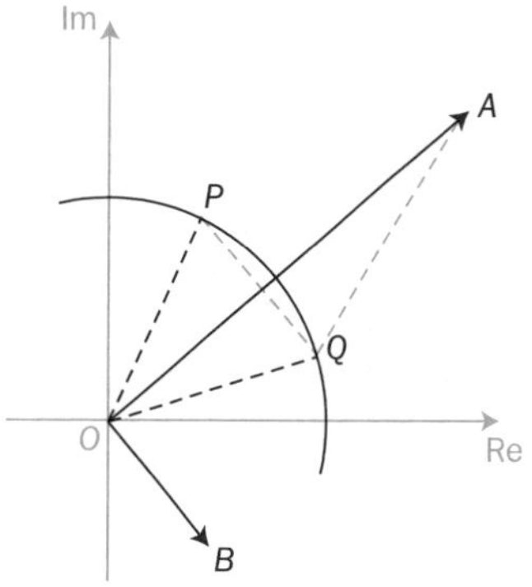

Exercise 2.6

1 **a** $|z|=7.6$, $\arg z=1.2^c$
b $|z|=6.4$, $\arg z=2.2^c$
c $|z|=6.3$, $\arg z=-0.3^c$
d $|z|=3.2$, $\arg z=-1.9^c$
e $|z|=1.0$, $\arg z=-0.7^c$
f $|z|=4.4$, $\arg z=1.9^c$

2 **a** 5, 0.9 **b** 13, 2.0 **c** 14.5, −0.8
d 7.5, −2.2 **e** $\frac{1}{3}$, 0.8 **f** 3, 2.2

3 **a** $|z| = 2\sqrt{3}$, $\arg z = \frac{1}{6}\pi$

b $|z| = 2\sqrt{2}$, $\arg z = \frac{3}{4}\pi$

c $|z| = 2\sqrt{2}$, $\arg z = -\frac{2}{3}\pi$

d $|z| = \sqrt{3}$, $\arg z = -\frac{1}{3}\pi$

e $|z| = 2\sqrt{5}$, $\arg z = \frac{\pi}{3}$

f $|z| = 1$, $\arg z = \frac{3\pi}{4}$

g $|z| = \sqrt{2}$, $\arg z = 0^c$

h $|z| = 3\sqrt{2}$, $\arg z = -\frac{\pi}{2}$

4 **a** 6.40 **b** -0.98^c **c** 7
d 2.03^c **e** 13.15 **f** 1.65^c

5 **a** $a = -4$ **b** 2.5^c

6 **a** $b = -2\sqrt{3}$ **b** 4

7 **a** $|z| = \frac{1}{2}$, $\arg z = \frac{1}{3}\pi$

b $|z| = \frac{3}{4}\sqrt{2}$, $\arg z = -\frac{1}{4}\pi$

c $|z| = \sqrt{2}$, $\arg z = -\frac{3}{4}\pi$

d $|z| = 1$, $\arg z = -\frac{1}{4}\pi$

e $|z| = \frac{1}{2}$, $\arg z = \frac{5}{6}\pi$

f $|z| = \frac{3}{2}$, $\arg z = -\frac{2}{3}\pi$

8 **b** $\arg z = -2.03^c$ **c** $|z| = -\alpha\sqrt{5}$

10 **a** $\sqrt{34}$ **b** $\sqrt{93}$

11 **a** $\arg z = \frac{1}{3}\pi$, $\arg w = \frac{5}{6}\pi$ **c** 4

12 **b** 0.68^c

Exercise 2.7

1 **a** $z = \sqrt{10}\,(\cos(1.25^c) + i\sin(1.25^c))$

b $z = \sqrt{29}\,(\cos(-1.19^c) + i\sin(-1.19^c))$

c $z = \frac{\sqrt{13}}{6}\,(\cos(2.16^c) + i\sin(2.16^c))$

d $z = 2\sqrt{5}\,(\cos(-2.03^c) + i\sin(-2.03^c))$

e $z = \sqrt{11}\,(\cos(1.13^c) + i\sin(1.13^c))$

f $z = \frac{\sqrt{30}}{6}\,(\cos(-0.68^c) + i\sin(-0.68^c))$

2 **a** $z = 0.21 + 2.99i$ **b** $w = -5.65 - 2.01i$
c $u = -0.39 + 1.45i$ **d** $v = 0.45 - 3.13i$

3

Modulus-argument form	Cartesian form
$z = 4\left(\cos\left(\frac{1}{3}\pi\right) + i\sin\left(\frac{1}{3}\pi\right)\right)$	$z = 2 + 2\sqrt{3}i$
$w = 4\sqrt{2}\left(\cos\left(\frac{3}{4}\pi\right) + i\sin\left(\frac{3}{4}\pi\right)\right)$	$w = -4 + 4i$
$p = \sqrt{6}\left(\cos\left(-\frac{1}{6}\pi\right) + i\sin\left(-\frac{1}{6}\pi\right)\right)$	$p = \frac{3\sqrt{2}}{2} - \frac{\sqrt{6}}{2}i$
$q = 2\left(\cos\left(-\frac{5\pi}{6}\right) + i\sin\left(-\frac{5\pi}{6}\right)\right)$	$q = -\sqrt{3} - i$

4 **a** $|z| = \sqrt{2}$, $|w| = 2$ **b** **i** $4\sqrt{2}$ **ii** $\sqrt{2}$
c $|z^2 - zw| = \sqrt{2}(1 + \sqrt{3})$

5 **b** **i** $\frac{1}{2}\sqrt{2}\left(\cos\left(-\frac{1}{4}\pi\right) + i\sin\left(-\frac{1}{4}\pi\right)\right)$

ii $\frac{1}{2} - \frac{1}{2}i$

6 **a** The given form has a subtract sign in the bracket

b $z = \frac{\sqrt{3}}{2} - \frac{3}{2}i$

c $z = \sqrt{3}\left(\cos\left(-\frac{1}{3}\pi\right) + i\sin\left(-\frac{1}{3}\pi\right)\right)$

7 **a** $|w| = 8\left(\cos\left(-\frac{\pi}{3}\right) + i\sin\left(-\frac{\pi}{3}\right)\right)$

b

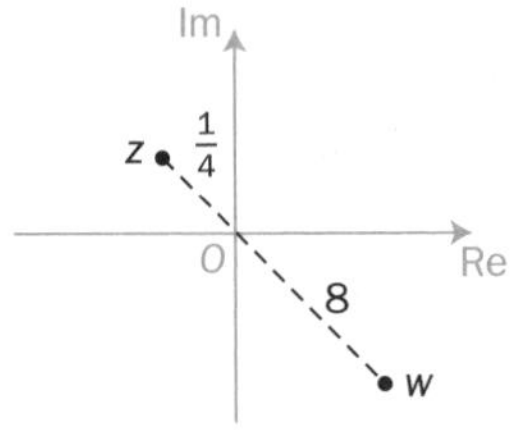

c $\frac{31}{4}$

8 **a** $\left|\frac{wu}{z}\right| = 5$ **b** $\frac{1}{4}\pi$

c

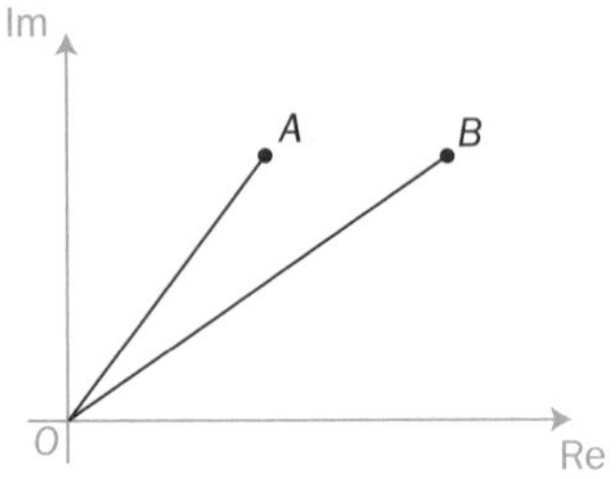

d Area $= \frac{1}{2}$

9 **a** $w = 2 - 2i$ **b** $v = -\sqrt{3} + i$

10 **a** 6 **b** $6\sqrt{2}$ **c** $\frac{1}{9}\sqrt{2}$ **d** $4\sqrt{2}$

11 **a** $4\sqrt{2}$ **b** $4 + 4\sqrt{3}$

12 **a** $z = 2\left(\cos\left(\frac{1}{6}\pi\right) + i\sin\left(\frac{1}{6}\pi\right)\right)$

$w = 3\left(\cos\left(\frac{3}{4}\pi\right) + i\sin\left(\frac{3}{4}\pi\right)\right)$

c **i** $\frac{9}{8}$ **ii** 2.17

13 **a** $|zw - z| = \sqrt{8}$, $|w - 1| = 2$
b $w = (\sqrt{3} + 1) + i$

14 $w = 5\left(\cos\left(\frac{11}{12}\pi\right) + i\sin\left(\frac{11}{12}\pi\right)\right)$

Exercise 2.8

1 **a** $0 + i$ **b** $-3 + 2i$
c $\frac{3}{4} - \frac{1}{2}i$ **d** $-1 + \sqrt{2}i$

2 **a** $-19 + 9i$ **b** $-15 + 30i$
c $-3 - 6i$ **d** $8 + 2i$

4 **a** $a = -4$ **b** $a = 9, b = -9$

5 **a** $b = 13$ **b** $k = -3$

6 $a = 2, b = 5$

7 **a** $h = 4$ **b** $z = -i$

8 **b** $w = \frac{1}{2} + \frac{1}{2}i$

9 **b** $z = \pm i$, $z = \pm\sqrt{5}i$

Exercise 2.9

1 **a** $(z - 1)$ **b** $(z + 2)$ **c** $(2z - 1)$

2 **a** $1 + 2i$

b $P(z) = (z - (1 + 2i))(z - (1 - 2i))(z + 1)$

3 **a** $(z - (3 + i)), (z - (3 - i)), (z - 1)$

b $(z - (2 - i)), (z - (2 + i)), (2z - 1)$

c $(z - (2 + 3i)), (z - (2 - 3i)), (z + 4)$

4 **a** $x = 4 \pm i$, $x = -\frac{1}{2}$ **b** $x = 2 \pm 2i$, $x = -4$

5 **a** $-2i$

b $P(z) = (z - 2i)(z + 2i)(z + 1)(z - 4)$

$z = \pm 2i$, $z = -1$, $z = 4$

6 **a** $x = 1 \pm 2i$, $x = \pm 2$ **b** $x = 2 \pm i$, $x = \pm 2i$

c $x = 2 \pm 3i$, $x = -1$, $x = -2$

d $x = -3 \pm i$, $x = 3 \pm i$

7 **a** $(1 + 2i)^3 = -11 - 2i$ **b** $z = 1 \pm 2i$, $z = -2$

8 **a** $h = 4$, $k = 6$

b $P(z) = (z + \sqrt{2}i)(z - \sqrt{2}i)(2z + 3)$

9 **a** B represents $1 - 2i$

C represents $5 + 0i$

c Kite. Area = 10

10 **a** $z = 2 + i$, $z = -4$

b

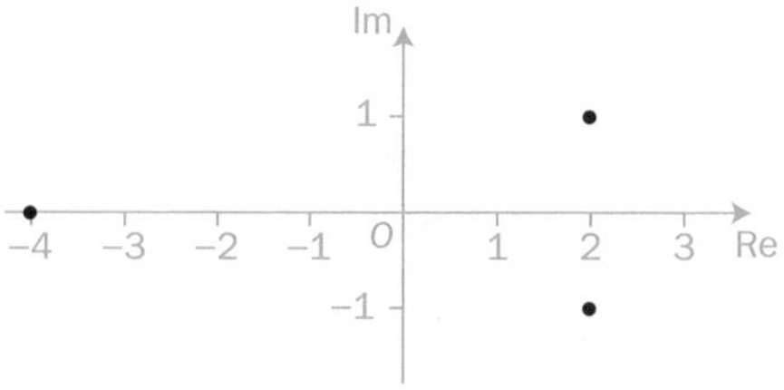

11 **a** $1 - i$

b $x = 1 \pm i$, $x = \pm i\sqrt{2}$

c

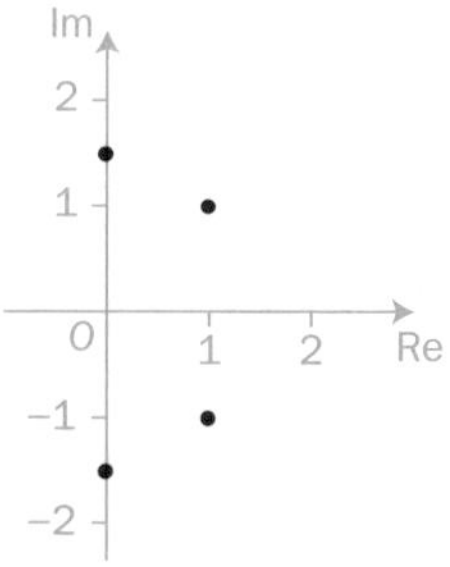

d Isosceles trapezium

Area $= 1 + \sqrt{2}$

13 **b** **i** e.g $P(x) = (x - 1)(x^2 + 1)$

ii e.g $P(x) = (x - 2i)(x^2 + 1)$

Review 2

1 **a** $z = \frac{5}{4} \pm \frac{\sqrt{7}}{4}i$ **b** $z = 1 \pm \frac{\sqrt{6}}{3}i$

c $z = -\frac{1}{8} \pm \frac{3\sqrt{7}}{8}i$ **d** $z = 1 \pm i$

2 **a** **i** $11i$ **ii** $-\frac{1}{26} + \frac{5}{26}i$

iii $16 + 30i$ **iv** $-\frac{2}{25} + \frac{6}{25}i$

b $\arg(z - w^2) = \frac{1}{2}\pi$

3 $z = \frac{1}{3} + \frac{7}{3}i$

4 **a** $\text{Im}(z) = \frac{\lambda + 1}{\lambda - 1}$

b **i** $\arg z = \frac{1}{2}\pi$ **ii** $\arg z = -\frac{1}{2}\pi$

5 $z = \frac{1}{2} + \frac{1}{2}i$

6 **a** $z = 6\sqrt{3} + 6i$

b

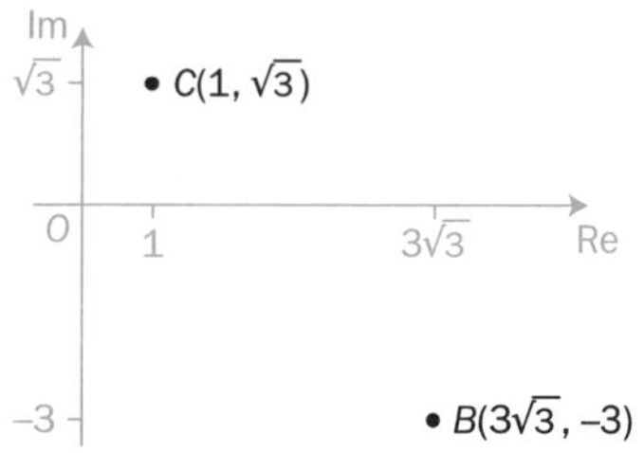

c $BC = 2\sqrt{10}$

7 **a** $|z| = 2\sqrt{2}$, $\arg z = \frac{1}{4}\pi$

b $-\sqrt{6} + i\sqrt{2}$

8 **a** $z = \frac{1}{4}k(1 + i\sqrt{3})$ **b** $\arg z = -\frac{2}{3}\pi$

c $z^3 = -\frac{1}{8}k^3 > 0$

9 **a** $2 - i$ **b** $z = 2 \pm i$, $z = -\frac{3}{2}$

10 **b** $z = 1 \pm 2i$, $z = -2$

11 **a** $z = 2 - 2\sqrt{3}i$, $z = -4$

b

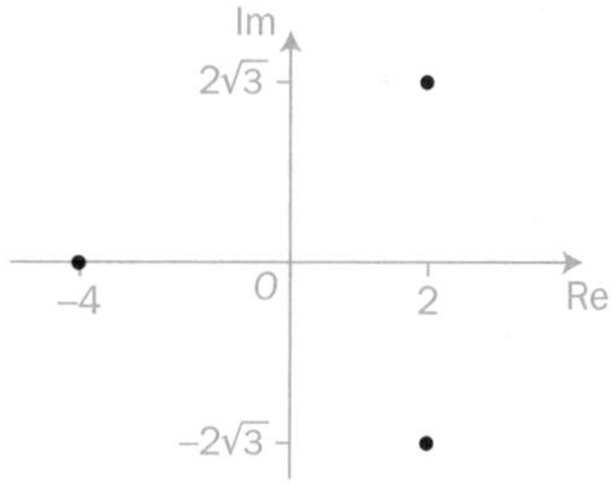

c radius = 4

d When added head to tail the vectors form an equilateral triangle.

12 **a** $x = -1 + 2i$, $x = -3$; A represents $-1 - 2i$, B represents $-1 + 2i$, C represents $-3 + 0i$

c $x^4 + 4x^3 + 6x^2 + 4x - 15 = 0$

13 **a** $w = 1 + 2i$ **b** 0.18^c

14 **a** $z = -1 \pm 4i$

b

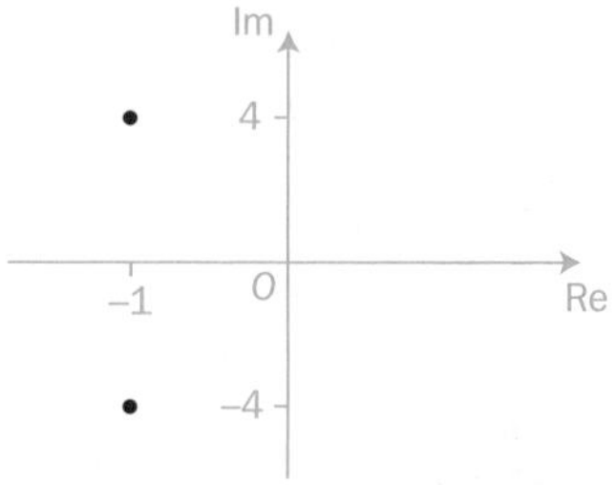

15 **a** $z = -1 + 2i$ **b** 2.03^c

16 **b** $a^2 - b^2 + 16 = 0$

$z = \pm(3 + 5i)$

17 **a** $|wz^2| = 16$

b

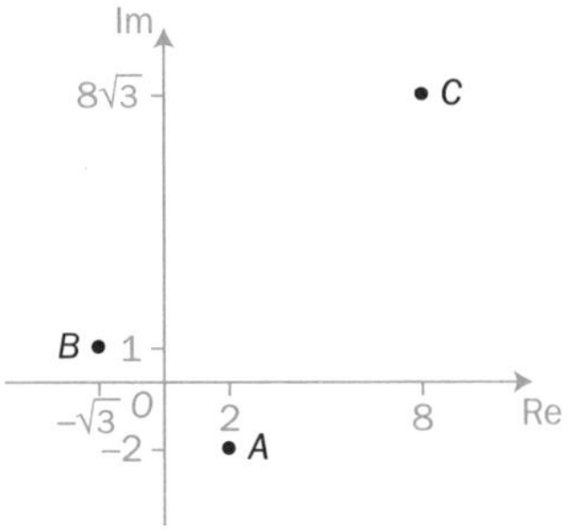

$B\hat{O}C = \frac{1}{2}\pi$

18 **a** $x = 3 \pm 2i, x = \pm i\sqrt{6}$

b

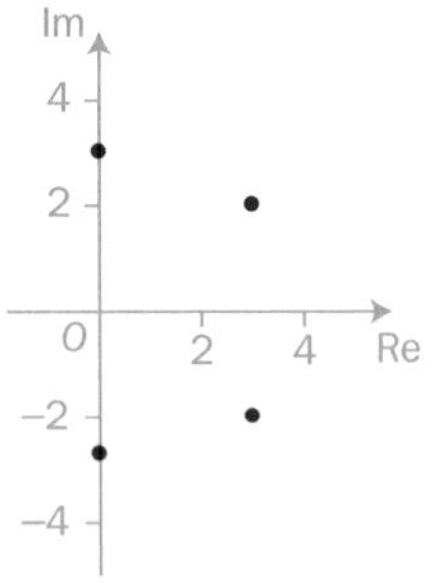

Chapter 3

Before you start

1 **a** $y = 3x + 12$ **b** $P(10, 0)$

2 **a** $f'(x) = 9x^{\frac{1}{2}} - 5$ **b** $f'(x) = 1 + x^{-2}$

c $f'(x) = -x^{-2} + 2x^{-\frac{3}{2}}$

3 **a** $x_1 = 1.5, x_2 = 0.5$ **b** $x_1 = 15, x_2 = 15$

Exercise 3.1

3 [0.7075, 0.7085); width is 0.001

4 **b** The principle assumes a change of sign has occurred. In this case $f(x) \geqslant 0$ with a minimum at $\left(\frac{1}{2}, 0\right)$.

Exercise 3.2

1 Answers are to 1 decimal place

a $\alpha = 1.3$ **b** $\alpha = 0.8$ **c** $\alpha = 5.4$ **d** $\alpha = -1.3$

2 **b** $\alpha = 0.64$ to 2 decimal places

3 Answers are to 1 decimal place

a $\alpha = 2.8$ **b** $\alpha = 0.8$

4 **b** [5.475, 5.4875]

5 **a**

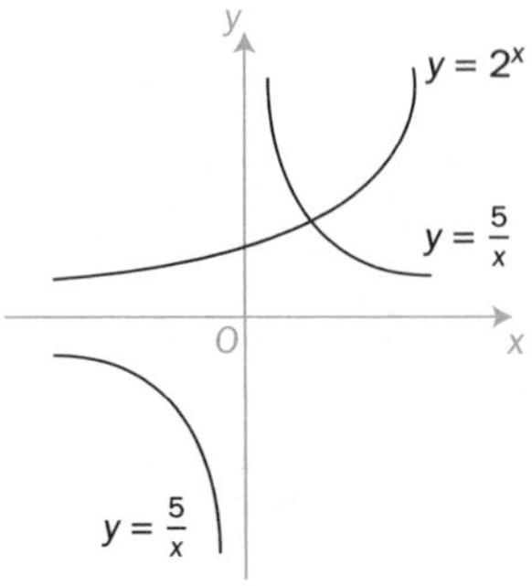

c $\alpha = 1.6$ to 1 decimal place.

6 **a**

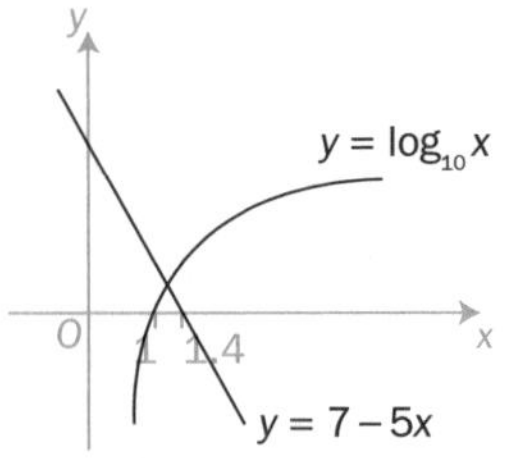

c $\alpha = 1.4$ (1 d.p)

7 **b** [2.225, 2.25]

$\alpha = 2.2$ to 1 decimal place.

8 **a**

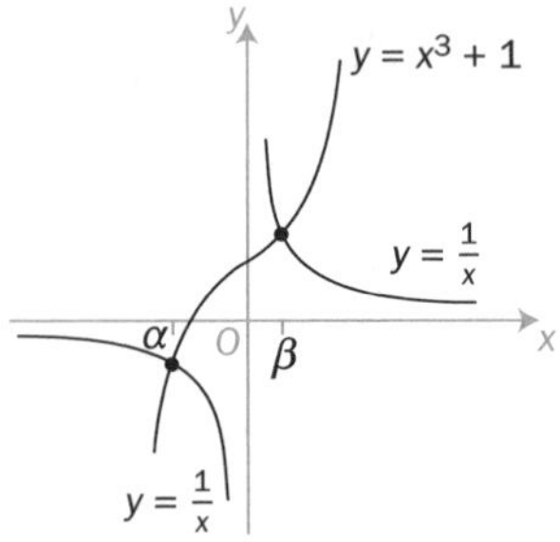

b (−1.25, −1) **d** $\beta = 0.7$

e f(−1) = 1, f(1) = 1

No change of sign (due to discontinuity of $y = f(x)$ over [−1, 1]).

9 **a** α

b No initial change of sign stops the method from working.

10 $\sqrt[3]{500} = 7.94$ to 2 decimal places.

11 **b** $\alpha = 2.9$ to 1 decimal place.

c The graph of $y = f(x)$ has an asymptote at $x = \pi$ and hence $f(x)$ is not a continuous function.

12 **a**

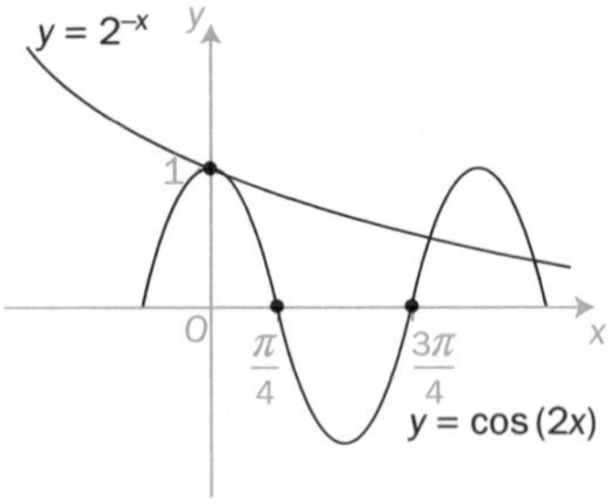

c $\alpha = 0.32$ to 2 decimal places

Exercise 3.3

1 **a** 1.5 **b** 1.6 **c** 0.125 **d** 1.25

2 **b** 1.25 (to 2 decimal places)

3 **a**

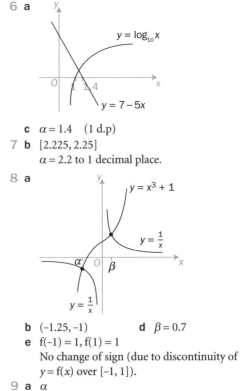

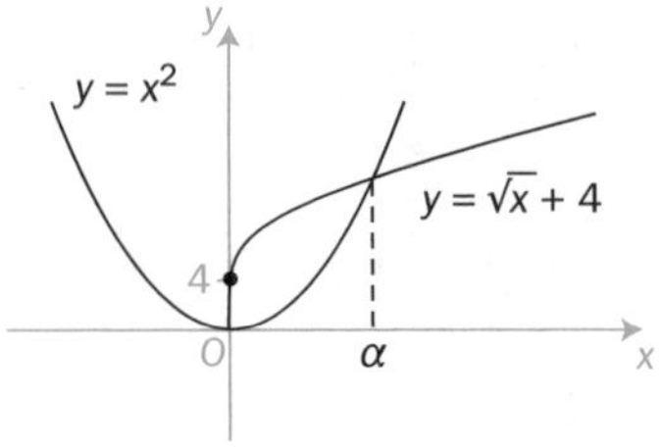

c 2.35 (to 2 decimal places)

4 **b** 1.75 (to 2 decimal places) **c** under-estimate

FP1

5 **a** Due the curvature of the graph, the straight line joining the given end-points crosses the x-axis at a value greater than α

b i For example,

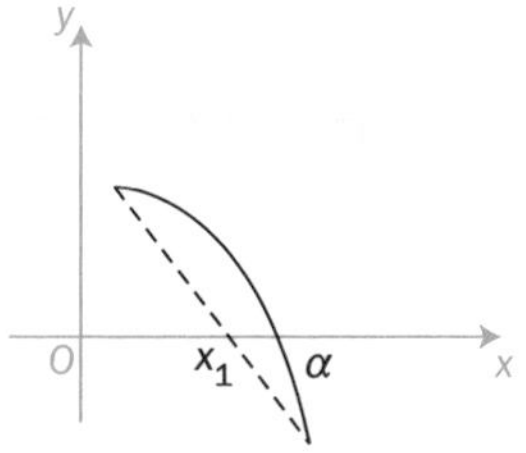

ii For example,

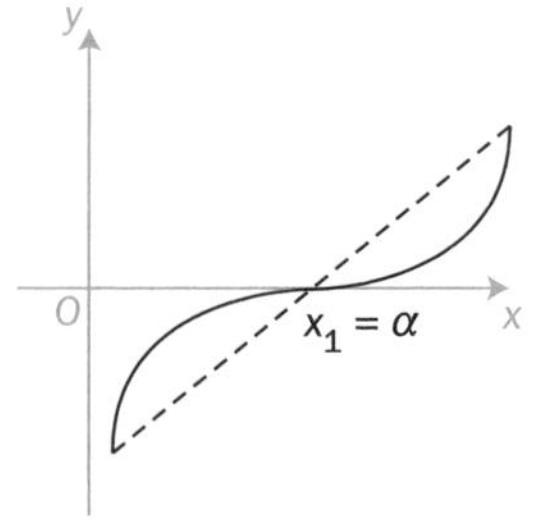

6 **a**

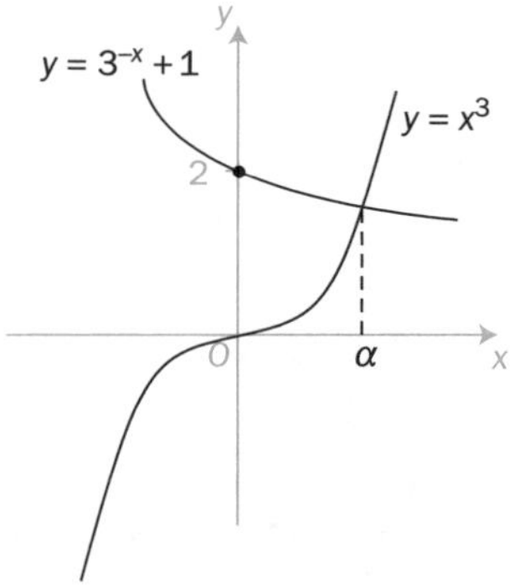

c 1.1 (to 1 decimal place)

d 1.09 (to 2 decimal places)

Exercise 3.4

1 Answers to 2 decimal places where appropriate

a 1.875 **b** 1.5 **c** −1.83 **d** −1.6 **e** 2.15

2 Answers to 3 decimal places

a 3.036 **b** 1.647 **c** 3.652

3 **a** $x_2 = \frac{9}{11} = 0.82$ (2 decimal places)

4 **a** 2.15 **b** $\alpha = \sqrt{\frac{3+\sqrt{37}}{2}}$ **c** 0.9%

5 **a** $x_2 = 3.944$ (3 decimal places)

$x_3 = 3.942$ (3 decimal places)

c i $x_2 = 0$ ii Root γ

6 **a** $f'(x) = -\frac{1}{x^2} + \frac{2}{x^3} + 1$

b $x_2 = 0.62$ (2 decimal places)

c Starting with $x_1 = 1$ produces $x_2 = 0.5$, the starting value for part **b**

7 **a** 2.3 (1 decimal place)

b $x_1 = 1$, $x_2 = -1$ which leads to an error on the next iteration

Starting near a stationary point has produced a negative iterate. The curve is not defined for $x \leqslant 0$

8 **b** 1.810457…

d $f'(1) = 0$

$x_1 = 1$ corresponds to a stationary point of the curve $y = f(x)$

9 **b** 1.90 (2 decimal places)

c i $\theta = \frac{1}{3}\pi$ ii $\theta_1 = -7.47\ldots$

iii The curve $y = f(\theta)$ has a stationary point at $\theta = \frac{1}{3}\pi \approx 1$

Review 3

1 **b** [1.45, 1.5]

c $\alpha = 1.5$ (to 1 decimal place)

2 **a** 2.95 (to 2 decimal places)

b The graph curves in such a way that the line joining the end-points cuts the x-axis with a value which is less than α

c $\alpha = 3.0$ to 1 decimal place

3 **b** $x_2 = 1.946$ (3 decimal places)

4 **a** $\alpha = 4.3$ (to 1 decimal place)

c The line joining the end-points crosses the x-axis outside the interval (4, 5)

5 **b** $\alpha = 3.65$ (to 2 decimal places)

c $x_2 = 3.66$ (to 2 decimal places)

6 **a** [0.9, 0.95]

$\alpha = 0.9$ (to 1 decimal place)

b 0.95 (to 2 decimal places)

7 **b** $x_2 = -1.947$ (to 3 decimal places)

c $x_2 = 6$

The starting value $x_1 = -1$ is near a stationary point $[f'(x) = 6x^2 - 2x - 9 = 0 \Rightarrow x = 1.402\ldots,$ $x = -1.069\ldots]$

8 **a** 2.35 (3 sig. figs) **b** $x_2 = 2.3981\ldots$

9 **a**

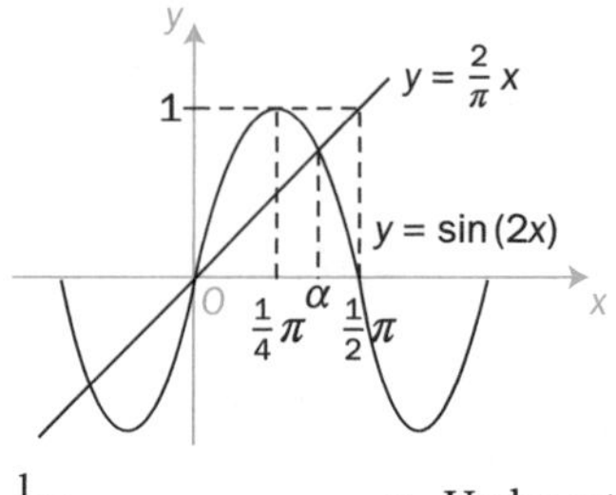

b $\frac{1}{3}\pi$ **c** Under-estimate

10 **b** 6.52 (to 2 decimal places)

c $x_1 = 4$ is the x-coordinate of a stationary point.

$f'(x) = 2x - 6x^{\frac{1}{2}} + 4 \Rightarrow f'(4) = 0$

11 **a**

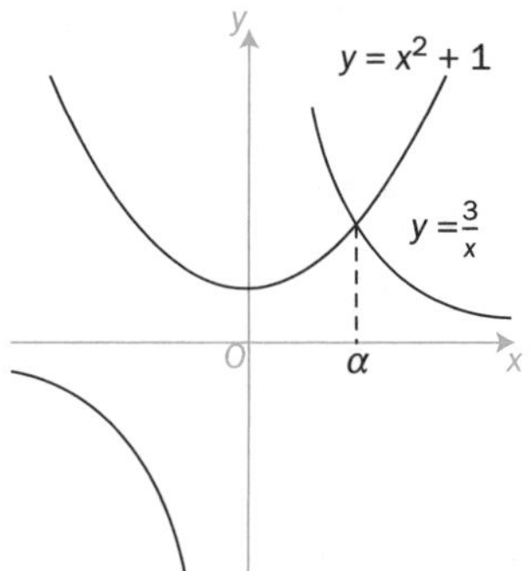

FP1

b $[1.21, 1.215]$
$\alpha = 1.21$ to 2 decimal places
c $x_2 = 1.21338\ldots$

12 **a** $x_2 = -0.5$
Root γ
b $x = \sqrt[3]{2} = 1.2599\ldots$ **c** Root β

Chapter 4

Before you start

1 **a** $(2, -3), (4, 1)$ **b** $\left(-\frac{1}{3}, -3\right), (1, 1)$
c $(0, 1), (-2, -1)$ **d** $(-2, 3), (-3, 2)$

2 **a** $y = 7x - 5, 7y + x = 15$
b $y = \frac{3}{2}x + 2, 3y + 2x = 32$
c $y = 5x + 2, 5y + x + 16 = 0$

Exercise 4.1

1 **a** $F(3, 0), x = -3$
b $F(2.5, 0), x = -2.5$
c $F\left(\frac{1}{8}, 0\right), x = -\frac{1}{8}$
d $F(\sqrt{3}, 0), x = -\sqrt{3}$

2 **b** 1.5
c $x = 2$ for $y^2 = 9x$
$x = 6$ for $y^2 = 3x$
d

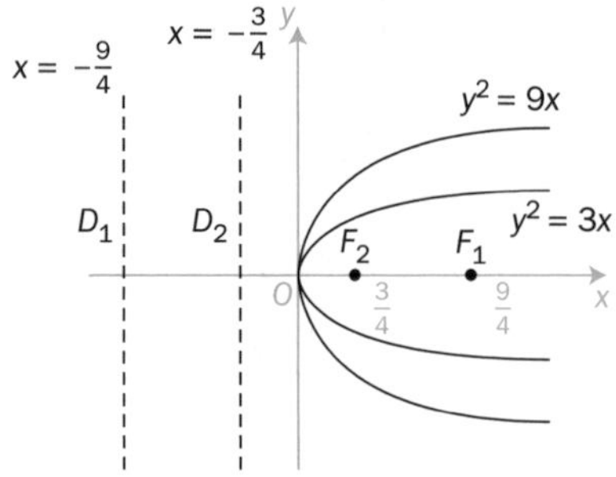

3 **a** $k = 9$ **b** $F\left(\frac{9}{4}, 0\right), x = -\frac{9}{4}$

4 **a**

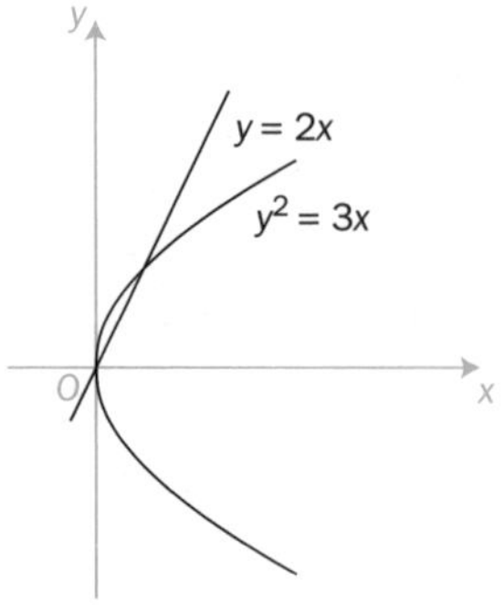

b $P\left(\frac{3}{4}, \frac{3}{2}\right)$ **c** $\frac{9}{16}$

5 **a** $k = \frac{1}{2}$ **c** $Q(8, -8)$

6 **a** $P\left(\frac{1}{4}a, a\right)$ **c** $Q\left(-a, \frac{8}{3}a\right)$

7 **a** $F(5, 0)$ **b** $x = 8$ **c** $y = \pm 4\sqrt{10}$

8 **b** $Q(-1, 2\sqrt{3})$ **c** $F(1, 0)$
d Triangle FPQ is equilateral. Area $= 4\sqrt{3}$

10 **a**

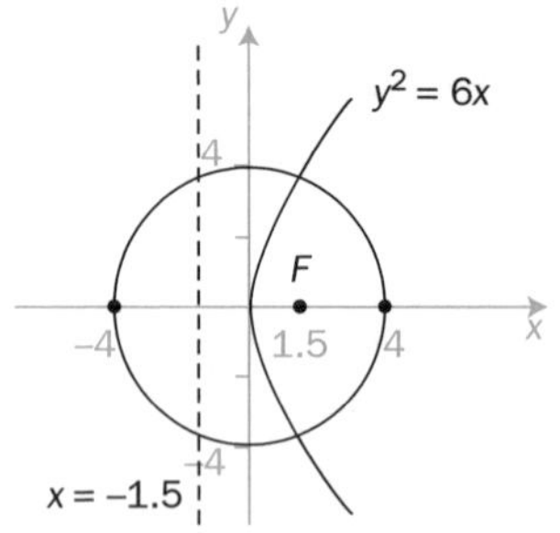

b $(2, \pm 2\sqrt{3})$

Exercise 4.2

1 **a**

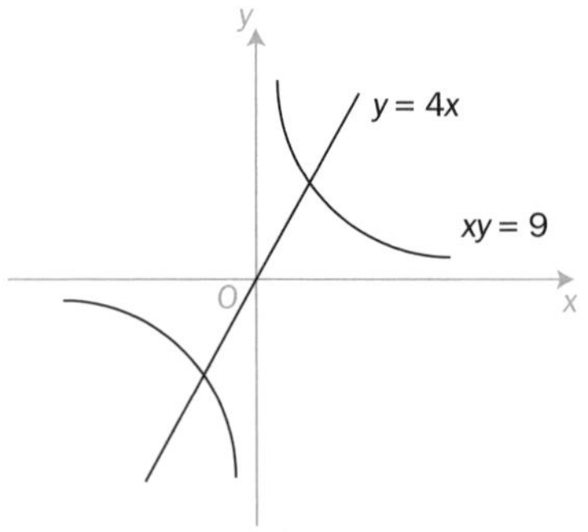

b $\left(\frac{3}{2}, 6\right), \left(-\frac{3}{2}, -6\right)$

2 Only points in parts **b** and **d** lie on the curve.

3 **a** $c = \sqrt{6}$ **b** $k = -8$

4 **a** $\left(-\frac{9}{2}, -8\right), (4, 9)$ **b** $(4, 4)$
c $(1, 7), (-1, -7)$

5 **b** The line with equation $y = 2c - x$ intersects C (it forms a tangent to the curve)

6 **b** $Q(3, 3)$ **c** $R(5, 0)$

7 **b** $(3, 4)$ or $(4, 3)$

8 **a** $P(1, \sqrt{8}), Q(\sqrt{8}, 1), R(-1, -\sqrt{8}), S(-\sqrt{8}, -1)$
b $y = \sqrt{8}x$ **c** $90°$

10 **a** The y-coordinates of P, Q and R are $2c$, c and $\frac{2}{3}c$ respectively.

11 **a** $c = 2\sqrt{2}$

Exercise 4.3

1 **a** $P\left(4t, \frac{4}{t}\right), t \in \mathbb{R}, t \neq 0$ **b** $P\left(\frac{3}{2}t^2, 3t\right), t \in \mathbb{R}$
c $P\left(\frac{1}{3}t, \frac{1}{3t}\right), t \in \mathbb{R}, t \neq 0$ **d** $P(\sqrt{2}t^2, 2\sqrt{2}t), t \in \mathbb{R}$

2 **a** $P(3t^2, 6t), t \in \mathbb{R}$
b

t	0	$\frac{1}{3}$	$\frac{1}{2}$	$\frac{\sqrt{2}}{2}$	2	$\frac{\sqrt{3}}{3}$
x	0	$\frac{1}{3}$	$\frac{3}{4}$	$\frac{3}{2}$	12	1
y	0	2	3	$3\sqrt{2}$	12	$2\sqrt{3}$

3 **b** $P\left(\frac{5}{4}t^2, \frac{5}{2}t\right), t \in \mathbb{R}$ **c** $t = -4$

4 **a** $P\left(12, \frac{3}{2}\right)$ **b** $t = -\sqrt{6}, a = -1, b = 6$
c $xy = 18$

5 a

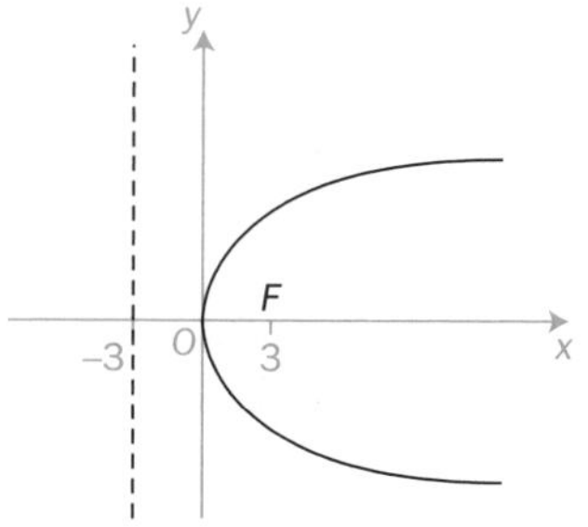

b $t = \pm\sqrt{2}$

6 a $P\left(\frac{1}{2}t, \frac{1}{2t}\right), t \in \mathbb{R}, t \neq 0$ **b** $12y = 12x + 5$

7 a $P\left(ct_0, \frac{c}{t_0}\right)$ **b i** $(-x_0, -y_0)$ **ii** (y_0, x_0)

8 b $t = \frac{1}{2}, t = -1: P\left(\frac{1}{2}c, 2c\right), Q(-c, -c)$

9 b $(1, 4), (16, -16)$

10 a $F(1, 0)$ **c** $OP^2 = t^4 + 4t^2$

d Area $= \sqrt{2} - 1$, $p = 2$, $q = -1$

11 b $t = \sqrt{2}$ **c i** $Q(2, 1)$ **ii** $k = 5$

Exercise 4.4

1 a $2y = x + 4, y = -2x + 12$

b $y = -16x + 16, 32y = 2x + 255$

c $2y = -3x - 4, 9y = 6x - 44$

d $2y = -x + 4\sqrt{3}, y = 2x - 3\sqrt{3}$

2 b $4y + 2x = 9$

3 b $3y = x + 16$ **c** $Q\left(-18, -\frac{2}{3}\right)$

4 a $k = 18$ **c** $Q\left(-\frac{9}{2}, \frac{15}{4}\right)$

5 a $P(4, 2), Q(-4, -2)$ **b** $(2, 4), (-2, -4)$

6 b $Q\left(\frac{27}{2}, 0\right)$ **c** Area $= 22.5$

7 b $Q(2x_0, 0), R(0, 2y_0)$

8 a $Q(-4, 0)$ **c** $P(4, 8)$

9 b $t^4y = 2t^2c - x$ **c** $R\left(\frac{2ct^2}{1+t}, \frac{2c}{t(1+t)}\right)$

10 a $\frac{1}{t}$ **c i** $Q\left(\frac{a}{t^2}, -\frac{2a}{t}\right)$

Review 4

1 a $y^2 = 8x$ **b** $\left(\frac{3}{2}, \pm 2\sqrt{3}\right)$

2 a $P(1, 10), Q(-5, -2)$

b No. The line with equation $y = 2x + 8$ is not normal to the curve at either P or Q.

3 c $Q(-8, -2)$

4 a $F\left(\frac{5}{2}, 0\right)$ **b** 5 **d** $\frac{25\sqrt{2}}{4}$

5 a Equation of tangent at P is $3y = x + 9$
Equation of tangent at Q is $3y = -x - 9$

b $k = -9$ **c** Area $= \frac{64}{3}$

d $(9t^2, 18t)$

6 a T_2 is $27y = -x - 36$ **b** $R(4.5, -1.5)$

c $\frac{1}{3}$ **d** Vertex P

7 a The gradient of the normal is negative.

b

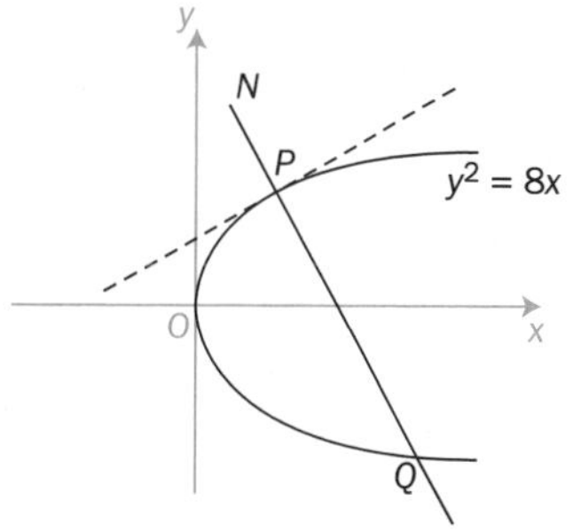

c $Q(16, -8\sqrt{2})$

d i $R(8, 0)$ **ii** $\cos P\hat{F}R = \frac{1}{3}$

8 a (c, c) **b** $-2c < k < 2c$

9 b $Q(x_0 + 2a, 0)$ **d** $x_0 = 3a$

11 b $4ty = x + 16at^2$ **c** $(4at^2, 5at)$

12 c $Q\left(-\frac{c}{t_0^3}, -ct_0^3\right)$

Revision 1

1 a $\frac{5+3p}{34} + \frac{5p-3}{34}i$ **b** $p = 4$

2 a $\left(\frac{3}{4}, 8\right), (-2, -3)$ **c** $D(-3, -7), xy = 21$

3 b 1.21 (3 sig. figs)

4 a $P(2, 4), Q(8, -8)$

5 a $2\sqrt{10}$

b

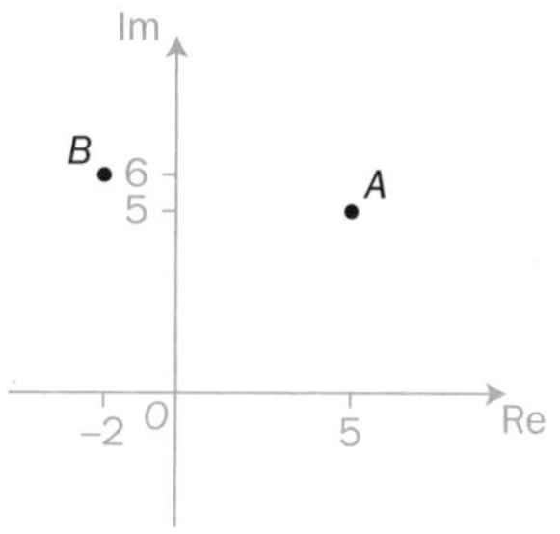

d -1.11^c (2 decimal places)

6 b $x = \pm 2i, x = \pm\frac{1}{2}i$

7 b $x = 3, x = -\frac{3}{2} \pm \frac{3}{2}\sqrt{3}i$

c

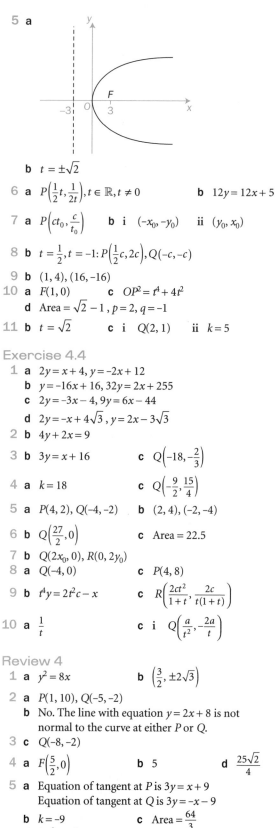

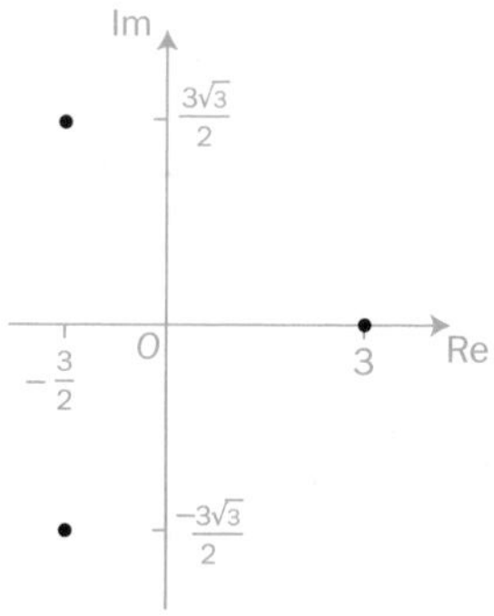

8 b $[1.3125, 1.325]$ or $(1.3125, 1.325)$

c 1.319 (3 decimal places)

9 a $F\left(\frac{3}{2}, 0\right)$ c −2 d $\sin Q\hat{P}F = \frac{1}{5}\sqrt{5}$

10 a -2.246^c (3 decimal places) b $\tan^{-1}(2)$

c $p = 5, w = \frac{1}{2} + i$

11 a 1.27 (2 decimal places) c 1.1881…

12 b $R\left(\sqrt{2}ct, \frac{c}{\sqrt{2}t}\right)$

13 a $x = 3 - i, x = \frac{1}{2}$ b $a = -13, b = 26$

14 b 6 c 20

15 b $t = \frac{1}{2}, t = -\frac{5}{4}$

$P(2, 8), Q\left(-5, -\frac{16}{5}\right)$

c $R(12.8, 1.25)$

16 b [1.855, 1.86] or (1.855, 1.86)

$\alpha = 1.86$ (2 decimal places)

c ii 1.8672…

Overestimate, since from part b, $1.855 < \alpha < 1.86$

d $f'(1) = 2 \times 1^{-\frac{1}{2}} - 2 \times 1 = 0$

$x_1 = 1$ is the x-coordinate of a stationary point on the graph of $y = f(x)$ and hence is not an appropriate starting value.

Chapter 5

Before you start

1 a i $Q(4, -4)$ ii $Q(4, 4)$ iii $Q(-4, -4)$ iv $Q(5, 5)$

2 a i $\frac{25}{4}\sqrt{3}$ ii $20\sqrt{2}$

b i $A(9, 0), B(6, 3)$ ii 13.5

Exercise 5.1

1 a 2×3 b 3×1 c 1×4 d 3×3

2 a 3 b 1 c −25

3 $\begin{pmatrix} 2 & 1 & 6 \\ 1 & 7 & \frac{3}{4} \\ 6 & \frac{3}{4} & 0 \end{pmatrix}$

4 $\begin{pmatrix} 0 & 1 & \pi & -\sqrt{2} \\ -1 & 0 & 2 & 0 \\ -\pi & -2 & 0 & -7 \\ \sqrt{2} & 0 & 7 & 0 \end{pmatrix}$

5 $\begin{pmatrix} \frac{1}{2} & \frac{1}{2} \\ \frac{1}{2} & \frac{1}{2} \end{pmatrix}$

Exercise 5.2

1 a $\begin{pmatrix} 5 & 2 & 3 \\ 12 & -2 & -5 \\ 15 & 9 & 11 \end{pmatrix}$ b $\begin{pmatrix} 2 & -10 \\ -\frac{1}{3} & -10 \\ -6 & 2 \end{pmatrix}$

c $\begin{pmatrix} 0 & 0 \\ 0 & -10 \end{pmatrix}$

2 a i $\begin{pmatrix} 3 & 8 \\ 6 & 1 \end{pmatrix}$ ii $\begin{pmatrix} -2 & -\frac{5}{2} \\ -3 & 3 \\ 1 & -2 \end{pmatrix}$ iii $k = -3$

b The matrices **A**, **B** and **C** are not all of the same order.

3 a $p = 4, q = 9, x = -2$ b $y = 0, y = 2$

4 a $\begin{pmatrix} 18 \\ 20 \end{pmatrix}$ b $\begin{pmatrix} -27 \\ 9 \end{pmatrix}$ c $(-2 \quad 6 \quad -8)$

d $\begin{pmatrix} 16 & 24 \\ 25 & 51 \end{pmatrix}$ e $\begin{pmatrix} 42 \\ 3 \\ 46 \end{pmatrix}$ f $\begin{pmatrix} 3 \\ 3 \\ 16 \end{pmatrix}$

5 a i $\begin{pmatrix} -7 & 4 \\ -4 & 0 \end{pmatrix}$ ii $\begin{pmatrix} -1 & 1 \\ 4 & 0 \end{pmatrix}$

b $\begin{pmatrix} 23 & -7 \\ 4 & -4 \end{pmatrix}$

6 a 3×1 b The product **BA** is not defined.

7 b $\mathbf{A}^3 = 3\mathbf{A}$ c $\mathbf{A}^5 = 9\mathbf{A}$

8 a i $3p + 2q = 5, 7p - 3q = 27$

ii $p = 3, q = -2$

b $x = 5.5, y = 3.5$

9 $x = -\frac{4}{3}, y = -2$

Exercise 5.3

1 a 3 b 14 c 2 d −2

2 a i $\begin{pmatrix} 4 & 3 \\ -5 & 3 \end{pmatrix}$ ii 27

c $\det A + \det B = 8 + 5 = 13 \neq 27$

3 a $x = 7$ b $x = \frac{1}{2}$ c $x = \pm 4$

d $x = -2, x = 3$ e $x = \frac{1}{2}, x = -\frac{4}{3}$ f $x = 1, x = 5$

4 a non-singular b singular

c singular d non-singular

5 b $y = \frac{3}{2}$

7 b $k = 2$ c $\det(\mathbf{AB}^2) = -40$

8 $p = 1, q = 17$

9 a $k = -\frac{7}{2}$ c $\lambda = \frac{1}{4}$

10 b $\frac{4}{9}$

11 b $x = 0, \pm 1$ c $x = \pm 2$

Exercise 5.4

1 a $\frac{1}{5}\begin{pmatrix} 2 & -3 \\ -1 & 4 \end{pmatrix}$ b $\frac{1}{20}\begin{pmatrix} 3 & -1 \\ 2 & 6 \end{pmatrix}$

c $-\frac{1}{25}\begin{pmatrix} 3 & -4 \\ -7 & 1 \end{pmatrix}$ d $\begin{pmatrix} 4 & -3 \\ -\frac{1}{3} & \frac{1}{2} \end{pmatrix}$

2 a $\begin{pmatrix} -2 & 3 \\ 3 & -4 \end{pmatrix}$ b $k = \pm 2\sqrt{2}$

3 **a** $\begin{pmatrix} 1 & -1 \\ -1 & 1.5 \end{pmatrix}$ **b** $\begin{pmatrix} 3 & 2 \\ 2 & 2 \end{pmatrix}$

4 **a** $-\frac{1}{15}\begin{pmatrix} -1 & -2 \\ -3 & 9 \end{pmatrix}$ **b** Not possible

c $\frac{1}{150}\begin{pmatrix} 8 & -2 \\ -81 & 39 \end{pmatrix}$

5 **a** $\frac{1}{12-2k}\begin{pmatrix} 3 & -k \\ -2 & 4 \end{pmatrix}$ **b** $k = 12$

6 **a** $\frac{1}{x^2+y^2}\begin{pmatrix} x & y \\ -y & x \end{pmatrix}$ **b** $\frac{1}{2}\begin{pmatrix} -1 & -1 \\ 1 & -1 \end{pmatrix}$

8 **a** $-\frac{1}{2}\begin{pmatrix} 2 & -2 \\ -7 & 6 \end{pmatrix}$ **c** $\begin{pmatrix} 2 & 3 \\ 1 & \frac{1}{2} \end{pmatrix}$

9 **a** $-\frac{1}{5}\begin{pmatrix} -4 & -3 \\ 13 & 11 \end{pmatrix}$ **b** $(\mathbf{AB})^{-1}\,\mathbf{A} = \mathbf{B}^{-1}$

$\mathbf{B}^{-1} = \begin{pmatrix} 3 & -1 \\ -10 & 4 \end{pmatrix}$

10 **b** The product **QP** is not defined and so $\mathbf{QP} \neq \mathbf{I}_{2\times 2}$

11 **b** $\frac{1}{2}\begin{pmatrix} 4 & -5 \\ -2 & 3 \end{pmatrix}$ **c ii** $x = \frac{3}{2}, y = -\frac{5}{2}$

d $x = 7, y = 2$

13 **c** $af + bh = 0, cf + dh = 1$

$f = -\frac{b}{ad-bc}, h = \frac{a}{ad-bc}$

Exercise 5.5

1 **a** $\begin{pmatrix} -1 & 0 \\ 0 & -1 \end{pmatrix}$ **b** $\begin{pmatrix} 0 & 1 \\ 1 & 0 \end{pmatrix}$ **c** $\begin{pmatrix} 0 & 1 \\ -1 & 0 \end{pmatrix}$

d $\begin{pmatrix} 0 & -1 \\ -1 & 0 \end{pmatrix}$ **e** $\begin{pmatrix} 1.5 & 0 \\ 0 & 1.5 \end{pmatrix}$

2 **b** $\begin{pmatrix} -\frac{1}{\sqrt{2}} \\ \frac{1}{\sqrt{2}} \end{pmatrix}$ **c** $\begin{pmatrix} \frac{1}{\sqrt{2}} & -\frac{1}{\sqrt{2}} \\ \frac{1}{\sqrt{2}} & \frac{1}{\sqrt{2}} \end{pmatrix}$

3 **a** $k = 3.5$ **b** $\begin{pmatrix} 3.5 & 0 \\ 0 & 3.5 \end{pmatrix}$

4 **b** Under this transformation, the origin is not fixed (it is mapped to the point (2, 2)) and so the transformation cannot be linear.

5 **a ii** $B(\sin\theta, \cos\theta)$ **b** $\begin{pmatrix} \cos\theta & \sin\theta \\ -\sin\theta & \cos\theta \end{pmatrix}$

6 **a** $\begin{pmatrix} -\frac{1}{\sqrt{2}} & -\frac{1}{\sqrt{2}} \\ \frac{1}{\sqrt{2}} & -\frac{1}{\sqrt{2}} \end{pmatrix}$ **b** $\theta = 315°$

7 **b** $\begin{pmatrix} -\frac{1}{2} & \frac{\sqrt{3}}{2} \\ \frac{\sqrt{3}}{2} & \frac{1}{2} \end{pmatrix}$

Exercise 5.6

1 **a** i $A: \begin{pmatrix} -1 & 0 \\ 0 & -1 \end{pmatrix}$, $B: \begin{pmatrix} -1 & 0 \\ 0 & 1 \end{pmatrix}$

ii $BA: \begin{pmatrix} 1 & 0 \\ 0 & -1 \end{pmatrix}$

b i $A: \begin{pmatrix} 1 & 0 \\ 0 & -1 \end{pmatrix}$, $B: \begin{pmatrix} 0 & -1 \\ 1 & 0 \end{pmatrix}$

ii $AB: \begin{pmatrix} 0 & 1 \\ 1 & 0 \end{pmatrix}$

c i $A: \begin{pmatrix} 5 & 0 \\ 0 & 5 \end{pmatrix}$, $B: \begin{pmatrix} 0 & -1 \\ -1 & 0 \end{pmatrix}$

ii $BA: \begin{pmatrix} 0 & -5 \\ -5 & 0 \end{pmatrix}$

d i $A: \begin{pmatrix} \frac{1}{\sqrt{2}} & -\frac{1}{\sqrt{2}} \\ \frac{1}{\sqrt{2}} & \frac{1}{\sqrt{2}} \end{pmatrix}$, $B: \begin{pmatrix} 0 & 1 \\ 1 & 0 \end{pmatrix}$

ii $BA: \frac{1}{\sqrt{2}}\begin{pmatrix} 1 & 1 \\ 1 & -1 \end{pmatrix}$

2 **a** $\mathbf{A} = \begin{pmatrix} -1 & 0 \\ 0 & 1 \end{pmatrix}$, $\mathbf{B} = \begin{pmatrix} 0 & -1 \\ -1 & 0 \end{pmatrix}$

b $\begin{pmatrix} 0 & 1 \\ -1 & 0 \end{pmatrix}$

c rotation 270° anticlockwise about O

3 **a** $\mathbf{P} = \begin{pmatrix} \frac{1}{\sqrt{2}} & -\frac{1}{\sqrt{2}} \\ \frac{1}{\sqrt{2}} & \frac{1}{\sqrt{2}} \end{pmatrix}$, $\mathbf{Q} = \begin{pmatrix} 0 & -1 \\ 1 & 0 \end{pmatrix}$

b Consecutive anti-clockwise rotations of 45° about O is equivalent to a single anti-clockwise rotation of 90°about O. In matrix terms this means **PP** = **Q**, or $\mathbf{P}^2 = \mathbf{Q}$

d **I**

4 **a** $\mathbf{P} = \begin{pmatrix} 0 & 1 \\ 1 & 0 \end{pmatrix}$, $\mathbf{Q} = \begin{pmatrix} 1 & 0 \\ 0 & -1 \end{pmatrix}$

b $\mathbf{PQ} = \begin{pmatrix} 0 & -1 \\ 1 & 0 \end{pmatrix}$; anticlockwise rotation of 90° centre O

c $n = 2$

5 **a** Clockwise rotation 90° about O

b Anti-clockwise rotation 90° about O

6 **a** The given transformation when composed with itself leaves every point unmoved.
Hence $\mathbf{A}^2 = \mathbf{I}$ and so $\mathbf{A}^{-1} = \mathbf{A}$

b $\begin{pmatrix} 0 & 1 \\ 1 & 0 \end{pmatrix}$

FP1

7 a $\begin{pmatrix} -1 & -1 \\ 1 & -1 \end{pmatrix}$

c An enlargement scale factor $\frac{1}{\sqrt{2}}$ centre the origin followed by a clockwise rotation of 135° about the origin.

8 a 3 b 4 c -1 d 1

9 a Area = 80 sq units. b Area = 1.25 sq units.

10 a Area = 24 sq units. b 30° or 150°

11 b The area scale factor is 0. The 'area' of a straight line is 0 sq units.

12 a ±4.5 b $x = 2, x = \frac{1}{2}$ c Area = $\frac{8}{9}$

13 a $\begin{pmatrix} \cos\theta & -\sin\theta \\ \sin\theta & \cos\theta \end{pmatrix}$

c i $\sin 2\theta \equiv 2\sin\theta\cos\theta$

ii $\tan 2\theta \equiv \frac{2\tan\theta}{1-\tan^2\theta}$

Review 5

1 b $\frac{1}{100}\begin{pmatrix} 5 & 26 \\ 5 & 6 \end{pmatrix}$

c **A** and **B** are not square matrices.

2 a $s = 5, t = -10$ b $\mathbf{A}^3 = 15\mathbf{A} - 50\mathbf{I}$

3 a $\frac{1}{8}\begin{pmatrix} -2 & 4 \\ -3 & 2 \end{pmatrix}$ c $\mathbf{B}^2 = \begin{pmatrix} 2 & 0 \\ 0 & 2 \end{pmatrix}$ $\mathbf{B}^{-1} = \begin{pmatrix} \frac{1}{2} & 1 \\ \frac{1}{4} & -\frac{1}{2} \end{pmatrix}$

4 a **I** b **BA** c **A**

5 a $\det \mathbf{H} = p^2 - q^2$ c $\begin{pmatrix} p+q & p+q \\ -(p+q) & -(p+q) \end{pmatrix}$

6 a i $\begin{pmatrix} -1 & 0 \\ 0 & 1 \end{pmatrix}$ ii $\begin{pmatrix} -1 & 0 \\ 0 & -1 \end{pmatrix}$ iii $\begin{pmatrix} 0 & -1 \\ -1 & 0 \end{pmatrix}$

b i (3, –2) ii (5, –4)

7 a $\begin{pmatrix} 0 & -1 \\ 1 & 0 \end{pmatrix}$ b Reflection in the line $y = x$

c $\begin{pmatrix} -1 & 0 \\ 0 & 1 \end{pmatrix}$ Reflection in the y-axis d $\begin{pmatrix} -1 & 0 \\ 0 & 1 \end{pmatrix}$

8 a $\det \mathbf{M} = 6$ b $k = 3$, (16, 15)

9 a 45° c $\frac{1}{\sqrt{2}}\begin{pmatrix} 1 & -1 \\ 1 & 1 \end{pmatrix}$

d $\mathbf{Q}^4\mathbf{P}^4 = -2\sqrt{2}\mathbf{P}$, $\lambda = -2\sqrt{2}$

10 a $x = -\frac{2}{3}$

11 a 45°

b 180°

$\mathbf{BA} = \begin{pmatrix} -1 & 0 \\ 0 & -1 \end{pmatrix}$

12 a $\begin{pmatrix} 0 & -1 \\ -1 & 0 \end{pmatrix}$

b Enlargement, scale factor $\frac{1}{2}$ from O c $B(-4, 4)$

13 a $p = -2, q = -6, r = 4$

14 a Area = 9 sq units b Black

c Area = $\frac{3}{2}$, White

15 b e.g. $\mathbf{C} = \begin{pmatrix} 3 & 1 \\ -1 & 3 \end{pmatrix}$ [accept any matrix **C** such that $c_{11} + 2c_{21} = 1$ and $c_{12} + 2c_{22} = 7$]

16 a 3 b $OB = 4$ units c 72 units^2

Chapter 6

Before you start

1 a $n(n+1)(n+3)$ b $n(n-1)(n+3)$

2 a 4060 b -1134

Exercise 6.1

1 a 30 b 420 c 155
d 1155 e 1285 f 797 040
g 18 496 h -1 344 760

2 b 4804

3 a 542 b 19 840 c 2847
d 9138 e 50 050 f 13 517 400

4 b $N = 29$

6 b 56 867 c $N = 3$

7 a 105 b 671 c 9315
d 10 880 e 2175 f 776 699

9 b 1 095 886

10 a $\frac{1}{4}n(n+1)(n-1)(n+2)$

11 a $n(n+1)(2n+1)(2n-1)$
b $n(2n+1)(7n+1)$
c $n^2(3n+1)(5n+3)$

13 a $T_n = \sum_{r=1}^{n}(n+1-r)r^2$

Review 6

1 a 1600 b 3621 c 844 200
d 27 378 e 27 110 f 32 344

2 b 120 140

3 a $n^2(n+1)$

4 b 5902

6 b $N = 5$

7 c $n(n+1)(3n-2)(n-1)$

10 b 105 488

Chapter 7

Before you start

1 a $2k(k+2)(k+1)$ b $4(k+1)$
c $k(k+2)(k+3)$ d $(k-1)^2(k+1)$

2 a $u_2 = \frac{1}{3}, u_3 = \frac{3}{4}, u_4 = \frac{4}{7}$

b i 26.5 (1 decimal place) ii 27

3 a i $\begin{pmatrix} 9 & -1 \\ 5 & 7 \end{pmatrix}$ ii $\begin{pmatrix} 0 & -28 \\ 7 & 21 \end{pmatrix}$ iii $\begin{pmatrix} 1 & -1.5 \\ -2 & 3.5 \end{pmatrix}$

b i $\begin{pmatrix} 1 & 0 \\ 0 & 4 \end{pmatrix}$ ii $\begin{pmatrix} 1 & 0 \\ 0 & 8 \end{pmatrix}$ iii $\begin{pmatrix} 1 & 0 \\ 0 & 2^n \end{pmatrix}$

c $\frac{8}{3}$

FP1

Exercise 7.1

1 **a** $S(k+1)$: "$5^{k+1} - 1$ is a multiple of 4"
b $S(k+1)$: "$2^{2k+1} + 1$ is exactly divisible by 3"
c $S(k+1)$: "$1 \times 2 + 2 \times 5 + 3 \times 8 + \cdots + (k+1)(3k+2) = (k+1)^2(k+2)$"

Review 7

4 **b** $N = 9$
5 **b** $\frac{20}{3}$
6 **a** $A = 2, B = 7, C = -4$
7 **b** 505
8 **c** $\begin{pmatrix} \left(\frac{1}{2}\right)^n & 0 \\ 0 & \left(\frac{1}{3}\right)^n \end{pmatrix}$, $p = \frac{1}{2}, q = \frac{1}{3}, b = c = 0$
10 **b** 417
11 **b** $\begin{pmatrix} 2n+1 & -2n \\ 2n & 1-2n \end{pmatrix}$
14 **c** ii $\begin{pmatrix} -6n+1 & 6n \\ -6n & 6n+1 \end{pmatrix}$
16 **a** $4(9^n - 2)$
17 **a** $\begin{pmatrix} 4 & 0 \\ -3 & 1 \end{pmatrix}$ **c** $a = 1024, N = 10$
19 **b** $\frac{16}{49}, q = \frac{4}{7}$
22 **b** $N = 1$

Revision 2

1 **a** $(2k+1)(2k+3)$
2 **b** 26851
3 $x = -1$
6 **b** 26660
7 **a** $\begin{pmatrix} 9+x & 8x \\ 8 & x+25 \end{pmatrix}$ **b** $k = 11$ **c** $\begin{pmatrix} 4 \\ 49 \end{pmatrix}$
9 **b** $\frac{1}{192}$
10 **a** $\mathbf{A} = \begin{pmatrix} 1 & 0 \\ 0 & -1 \end{pmatrix}, \mathbf{B} = \begin{pmatrix} 0 & 1 \\ 1 & 0 \end{pmatrix}$ **b** $\begin{pmatrix} 0 & -1 \\ 1 & 0 \end{pmatrix}$
c i Anti-clockwise rotation 90° about O
ii Rotation 180° about O
11 **b** $n^2(2n+3)$
13 **a** B
14 **b** $\frac{1}{6}n(n+1)(3n+2)(1-n)$
15 **b** ii $\mathbf{A}^3 = 39\mathbf{A} - 70\mathbf{I}, p = 39, q = -70$ **c** $\begin{pmatrix} 50 \\ 50 \end{pmatrix}$
16 **a** $\frac{6x+10}{x+3} \equiv 6 - \frac{8}{x+3}$, $p = 6, q = -8$
17 **b** $\det \mathbf{M} = 0.5$
i Area $= \frac{1}{4}k^2$
ii the area of the image ($2k^2$) exceeds that of the square OABC (k^2)
18 **b** $\mathbf{P}^{2n} - \mathbf{P}^n = 2n\begin{pmatrix} -2 & 1 \\ -4 & 2 \end{pmatrix}$, $\mathbf{A} = \begin{pmatrix} -2 & 1 \\ -4 & 2 \end{pmatrix}$
19 **b** $\begin{pmatrix} \frac{1}{6} & \frac{1}{6} \\ -\frac{1}{6} & \frac{1}{6} \end{pmatrix}$
$\mathbf{M}^{-1}$ represents a clockwise rotation through 45° about O followed by an enlargement scale factor $\frac{1}{3\sqrt{2}}$ centre the origin.
c i

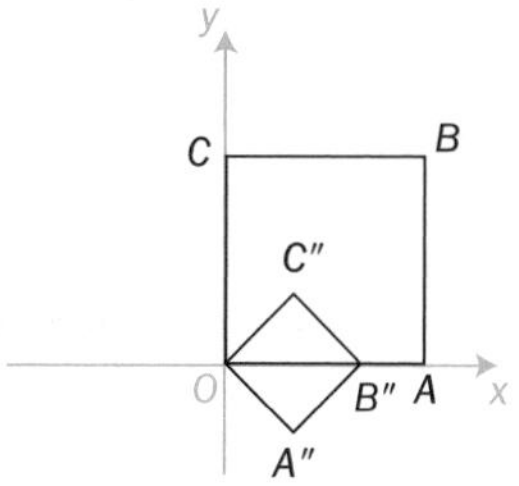

ii $\frac{1}{36}$
20 **b** i $f(n+1) - 3f(n) = 2u_n$

arcsin x If $\theta = \arcsin x$ then $x = \sin\theta$.

arg z The argument of the complex number z, usually denoted by θ.

Argand diagram An Argand diagram is a geometric representation of the complex number $z = x + iy$ by the point with co-ordinates (x, y). The horizontal axis is the real axis and the vertical axis is the imaginary axis.

Argument The argument of the complex number z is the angle between the positive real axis and the vector representing z on the Argand diagram.

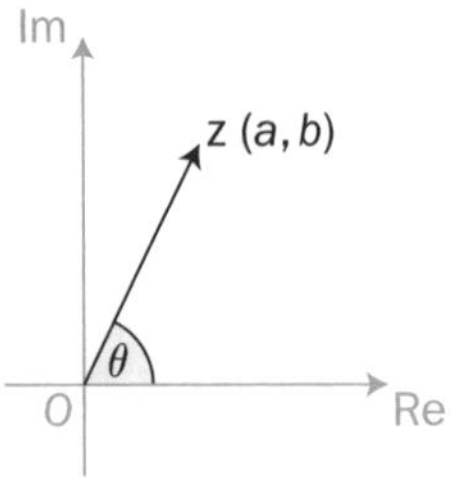

Arithmetic sequence A sequence generated by the same constant, or common difference, being added onto the previous term.
e.g. 2, 5, 8,.......

Arithmetic series A series generated by the same constant, or common difference, being added onto the previous term.
e.g. 2, 5, 8,.......

Asymptote A straight line which a curve approaches, gets nearer and nearer to, but never actually touches or crosses.

Bisect Cut into half.

Cartesian form Cartesian form of a complex number, z, is $z = a + ib$ where a, b are real numbers.

Change of sign method A method of determining whether a root of $f(x) = 0$ lies between two values by considering the sign of $f(x)$.

Closed interval $[a, b]$ If x lies in the closed interval $[a, b]$ then $a \leqslant x \leqslant b$.

Coefficient The constant used as a multiplier of a term including a variable.
e.g. 3 is the coefficient of x in the term $3x$ and 7 the coefficient of x^2y in the term $7x^2y$.

Column of a matrix In the matrix $\begin{pmatrix} 2 & 3 \\ 9 & 5 \end{pmatrix}$, $\begin{pmatrix} 2 \\ 9 \end{pmatrix}$ and $\begin{pmatrix} 3 \\ 5 \end{pmatrix}$ are columns of the matrix.

Compatible Two matrices are compatible if they can be multiplied, i.e. if the number of columns in the left-hand matrix is the same as the number of rows in the right hand matrix.

Completing the square Rearranging a function in the form $(ax + b)^2 + c$ where c is a constant.

Complex conjugate, z^* If $z = a + ib$ then its complex conjugate is $z^* = a - ib$.

Complex number An imaginary number in the form $a + ib$ where a and b are real and $i = \sqrt{-1}$.

Conjugate pairs $a + ib$ and $a - ib$ are conjugate pairs of complex numbers.

Continuous A function $f(x)$ is continuous if there are no breaks in the graph of $y = f(x)$.

Cosine rule In any triangle ABC $a^2 = b^2 + c^2 - 2bc\cos A$.

Cubic equation An equation which includes an x^3 term and no term of a higher power.

Denominator The 'bottom' of a fraction.
e.g. the 4 in $\frac{3}{4}$ or the $(x - 3)$ in $\frac{x}{x-3}$.

Derivative The result when a function has been differentiated.

Determinant of a matrix In a 2×2 matrix $\begin{pmatrix} a & b \\ c & d \end{pmatrix}$ the determinant is $ad - bc$. The determinant of a matrix A can be abbreviated to det A or $|A|$.

Differentiate To find the gradient of the tangent to a curve.

Directrix The directrix of a parabola $y^2 = 4ax$ is the line $x = -a$.

Discriminant In a quadratic equation $ax^2 + bx + c = 0$ the discriminant is $b^2 - 4ac$.

Divisor In division the term which is divided into the other.
e.g. in $6 \div 5$, 5 is the divisor, or in $\frac{x^2+6}{x-3}$, $(x - 3)$ is the divisor.

Elements of a matrix The numbers, or quantities, in a matrix.

Evaluate Work out the sum and give a single numeric answer.

Exact form A numerical value which is exact
e.g. 4π and $\sqrt{2}$.

Expansion The result after multiplying out bracket(s)

Exponential curves Curves with equations of the form $y = a^x$, where a is a positive constant.

Exponential graphs The graph of the curve $y = a^x$, where a is a positive constant.

Factor A factor is a number or an expression which divides exactly into another.

Finite series A series which has a finite number of terms.

Focus (plural foci) The focus of the parabola $y^2 = 4ax$ is the point $(a, 0)$

Focus-directrix property of a parabola The perpendicular distance from any point on a parabola to its directrix is the same as the distance from that point on the parabola to the focus.

Identity matrix, I A square matrix in which the elements of the leading diagonal all equal to 1 and all the remaining elements are zero.

Imaginary number, i An imaginary number is a number that cannot be shown on the number line.
e.g. $\sqrt{-21} = i\sqrt{21}$.

Imaginary part, Im(z) b, the coefficient of i, in the complex number $a + ib$.

Induction A form of proof of a hypothesis.

Inductive hypothesis The hypothesis to be tested using induction.

Integers, $\mathbb{Q}$ Positive and negative whole numbers, including zero.

Interval A defined section of the real number line.

Interval bisection A method of estimating the value of a root of an equation by bisecting an interval known to contain the root and then repeating the process on the smaller interval.

Inverse of a matrix $\mathbf{A}^{-1}$ is the inverse of square matrix $\mathbf{A}$ and is such that $\mathbf{A}^{-1}\mathbf{A} = \mathbf{I}$ and also $\mathbf{A}\mathbf{A}^{-1} = \mathbf{I}$.

Iterate To carry out a number of numerical repetitions.

Iterative formula A rule, or formula, which generates the next term in a sequence.

Iterative sequence The consecutive results when an iterative formula has been applied.

Leading diagonal The elements in a diagonal line from the top left element to the bottom right element of a square matrix.

Linear interpolation To find an approximate root of the equation $f(x) = 0$ by assuming the graph of $y = f(x)$ is a straight line in an interval where the graph crosses the x-axis.

Matrix (pl. matrices) A collection of numbers arranged in rows and columns in curved brackets
e.g. $\begin{pmatrix} 1 & 2 & 6 \\ 5 & -1 & 4 \end{pmatrix}$

Modulus of a complex number, $|z|$ If $z = x + iy$, the length of the line from (x, y) to the origin on the Argand diagram.
$|z| = \sqrt{x^2 + y^2}$.

Modulus-argument form A complex number $z = r(\cos\theta + i\sin\theta)$ where r = modulus and θ = argument of the complex number.

Multiple A multiple of a number is any whole number which is the product of the original number and another.

Natural numbers, $\mathbb{N}$ Whole positive numbers, not including zero.

Newton-Raphson An iterative method of solving $f(x) = 0$.

Non-singular matrix The determinant of the matrix is not zero.

Normal The normal to a point on a curve is the straight line perpendicular to the tangent at that point.

Obtuse angle An angle whose size is between 90° and 180°

Open interval, (a, b) If x lies in the open interval (a, b) then $a < x < b$.

Order of a matrix The order is (number of rows) × (number of columns)
e.g. a 4×3 matrix has 4 rows and 3 columns.

FP1

Parabola A shape/graph typified by $y^2 = 4ax$, where a is a positive constant.

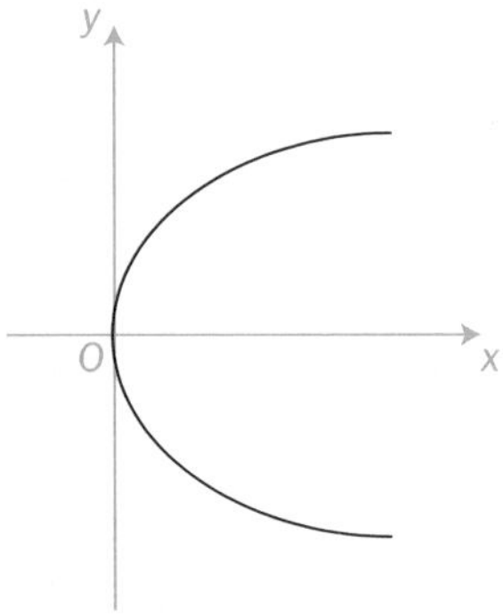

Parameter A curve can be defined by expressing x and y in terms of a third variable, called a parameter.

Parametric form Co-ordinates given in terms of a parameter.

Perpendicular lines Lines which are at right angles to each other.

Plane A surface with no depth.

Polynomial curves Curves with the equation $y = a_0 + a_1x + a_2x^2 + \ldots\ldots.. + a_nx^n$ where n is positive.

Polynomial equation An equation of the form $y = a_0 + a_1x + a_2x^2 + \ldots\ldots.. + a_nx^n$ where n is positive.

Prime numbers Numbers with no factors other than 1 and the number itself. 1 is **not** a prime number.

Principal argument The argument θ of a complex number when $-\pi < \theta \leqslant \pi$.

Product The result of a multiplication.

Purely imaginary A complex number is purely imaginary if it has no real part.
e.g. $2i$.

Pythagoras' Theorem In a right angled triangle the sum of the squares on the two shorter sides equals the square on the hypotenuse.

Quadrant One of the four parts formed in the plane by the x- and y-axes, or the real and imaginary axes.

Quadratic equation An equation with an x^2 term and no higher powers of x. Its general form is $ax^2 + bx + c = 0$ where a, b, and c are constant and $a \neq 0$

Quotient The answer when a function is divided by another.

Radians A unit of angle measurement. π radians = 180°, symbol ^{c}.

Real numbers, $\mathbb{R}$ Numbers which exists and can be shown on the number line.

Real part of a complex number, Re(z) The value of a in the complex number $a + ib$.

Rectangular hyperbola A curve with an equation of the form $xy = c^2$ where c is a non-zero constant.

Recurrence relation An iterative formula.

Roots The roots of an equation $f(x) = 0$ are the values of x which satisfy the equation.

Row of a matrix The horizontal elements in a matrix. In the matrix $\begin{pmatrix} 2 & 3 \\ 9 & 5 \end{pmatrix}$ $(2 \ \ 3)$ is a row and so is $(9 \ \ 5)$.

Self inverse matrix A matrix **A** is self inverse if $A = A^{-1}$.

Sequence A set of terms in which each consecutive term is derived from the previous one by following the same rule.
e.g 2, 5, 11, 2......

Series The sum of the terms in a sequence.

Simultaneous equations Two different equations in two variables which have a common solution.

Sine Rule In any triangle ABC, $\dfrac{a}{\sin A} = \dfrac{b}{\sin B} = \dfrac{c}{\sin C}$

Singular matrix The determinant of the matrix is zero.

Sketch A freehand drawing of a curve showing the shape of the curve and the points where the curve cuts the axes.

Square matrix A matrix with an equal number of rows and column.

Stationary point At this point the gradient of the function is zero. It can be a maximum or minimum point or a point of inflection.

Stretch A transformation where shape and position are changed.

Surd The square (or other) root of a number that produces an irrational number
e.g. $\sqrt{7}$.

Tangent A line which touches a curve and is parallel to the gradient of the curve at the point of contact.

FP1

Transformation An operation applied to an object (usually geometrical)

e.g. Translation, reflection, rotation, stretch etc.

Translation A transformation which moves an object or curve without changing its size or shape.

Trigonometric equations An equation involving trigonometric terms.

Trigonometric identity A relationship between trigonometric functions which is true for **all** values of the angle. The equivalence sign is represented by ≡.

Variable A letter which can take various numerical values, i.e. not a constant.

Vector A quantity with magnitude (size) and direction.

y-intercept The y-intercept is the y-coordinate of the point where a graph cuts the y-axis.

Zero matrix All the elements in the matrix are zero.

Formulae

The following formulae will be given to you in the exam formulae booklet.
You may also require those formulae listed under Core Mathematics C1 and C2.

Summations

$$\sum_{r=1}^{n} r^2 = \frac{1}{6}n(n+1)(2n+1)$$

$$\sum_{r=1}^{n} r^3 = \frac{1}{4}n^2(n+1)^2$$

Numerical solution of equations

The Newton-Raphson iteration for solving $f(x) = 0 : x_{n+1} = x_n - \frac{f(x_n)}{f'(x_n)}$

Coordinate geometry

The perpendicular distance from (h, k) to $ax + by + c = 0$ is $\frac{|ah + bk + c|}{\sqrt{a^2 + b^2}}$

The acute angle between lines with gradients m_1 and m_2 is $\arctan\left|\frac{m_1 - m_2}{1 + m_1 m_2}\right|$

Conics

	Parabola	Rectangular Hyperbola
Standard Form	$y^2 = 4ax$	$xy = c^2$
Parametric From	$(at^2, 2at)$	$\left(ct, \frac{c}{t}\right)$
Foci	$(a, 0)$	Not required
Directrices	$x = -a$	Not required

Matrix transformations

Anticlockwise roation throught θ about O: $\begin{pmatrix} \cos\theta & -\sin\theta \\ \sin\theta & \cos\theta \end{pmatrix}$

Reflection in the line $y = (\tan\theta)x$: $\begin{pmatrix} \cos 2\theta & \sin 2\theta \\ \sin 2\theta & -\cos 2\theta \end{pmatrix}$

Index

FP1